MATERIALS SCIENCE AND TECHNOLOGIES

PHOTONIC CRYSTALS

TECHNOLOGY, THEORY AND CHALLENGES

MATERIALS SCIENCE AND TECHNOLOGIES

Additional books in this series can be found on Nova's website under the Series tab.

Additional e-books in this series can be found on Nova's website under the e-book tab.

MATERIALS SCIENCE AND TECHNOLOGIES

PHOTONIC CRYSTALS

TECHNOLOGY, THEORY AND CHALLENGES

DONALD CLAUS

EDITOR

NOVINKA

New York

For permission to use material from this book please contact us:
Telephone 631-231-7269; Fax 631-231-8175
Web Site: http://www.novapublishers.com

Library of Congress Cataloging-in-Publication Data

ISBN: 978-1-62618-193-9

Published by Nova Science Publishers, Inc. † New York

CONTENTS

PREFACE

Photonic crystals are periodically structured electromagnetic media, possessing photonic band gaps, ie., ranges of frequency in which light cannot propagate through the structure. In this publication, the authors present current research in the study of the technology, theory and challenges of photonic crystals. Topics include photonic crystal tunable filters driven by an electrical control signal; recent patents on tunable photonic crystals; diamond photonics; design of high performance tunable reflectors by using 1D ternary photonic band gap structures; photonic crystal fiber nonlinear characterization and application to generate multi-wavelength brillouin fiber lasers; and a new multilayer structure for temperature stable performance in infrared optical filtering.

Chapter 1 - The field of quantum information processing (QIP) takes advantage of the properties of quantum mechanics to perform tasks that are classically challenging, including exponentially faster computational algorithms, secure communication, and complex physical and biological modeling. A number of recent theoretical and experimental advances have identified promising candidates for the implementation of QIP. Although the quest for an optimal physical system to encode quantum information is far from being over, the negatively charged nitrogen-vacancy color center in diamond (NV^-) presents significant advantages over its counterparts. This is the only system, reported so far, in which a single spin can be initialized, manipulated, optically read out, all at room temperature. Moreover, the storage time can amount to thousands of gate durations (2 ms for electon spins, up to 1 second for nuclear spins - the latter would enable more than 10^3 logic gates). Indeed, several groups demonstrated single and two-qubit operations with NV^- in diamond, thus, meeting all DiVincenzo conditions for a quantum bit, besides *scalability*. In this chapter the authors review photon-assisted interactions between distant NVs as a way to overcome the scalability problem in diamond NV quantum processing. This implies the ability to convert spin states in NV^{-} stationary qubits into photon states and vice versa, combined with a reliable transfer of photons between specific NVs [not quite – let's talk about this point]. This interaction can be greatly facilitated in an integrated quantum photonic architecture (including photonic crystal cavities and waveguides) for confining photons around NV centers and connecting between distant NV sites. Here the authors review the advancements in the field of diamond photonics in the last decade. This includes photonic structure design, various fabrication approaches towards nano-scale processing in a single crystal diamond substrate, and the state-of-the-art experimental demonstrations of single and integrated photonic elements.

Chapter 2 - This chapter describes one of the main applications of photonic crystal fibre (PCF) as a nonlinear gain medium to generate multi-wavelength fiber laser based on stimulated Brillouin scattering (SBS) and four-wave mixing (FWM) nonlinear effects. Firstly, the threshold power condition of SBS for a piece of 100 m long PCF has been theoretically studied in terms of fibre parameters and input pump power. The threshold exponential gain, G_{th}, can be anticipated by this simulation which strongly depends on the fibre length Brillouin gain content and effective area. This value varies from 14 to 18 depending on the fibre length according to the Smith's definition. In order to characterize completely, the frequency shift of Brillouin Stokes in the PCF is measured nearly 9.75 GHz via heterodyne measurement method. To demonstrate the PCF application in fiber laser generation process, a simple multi-wavelength Brillouin fibre laser (BFL) is illustrated using the PCF in conjunction with a fibre Bragg grating (FBG) to produce at least 7 Stokes and 4 anti-Stokes lines with a line spacing of 0.08 nm. To produce double-spacing Brillouin lines, a compact multi-wavelength Brillouin fibre laser in conjunction with a figure-of-eight configuration is proposed. At a BP power of 15.3 dBm, at least 4 simultaneous lines with 20 GHz or 0.16 nm line spacing is achieved by removing the odd-order Stokes lines. The anti-Stokes lines are also generated via FWM process in the laser cavity. The output spectrum of the proposed BFL can be tuned by 80 nm, dependent on the availability of an appropriate BP source. These results show that the use of a highly nonlinear PCF is capable to produce a compact and efficient multi-wavelength laser for many applications such as wavelength division multiplexing system (WDM) optical communication network system and sensors.

Chapter 3 – The authors are presenting a novel model of tunable filters based on photonic crystal waveguides and resonant cavities to work in high transmission rates optical systems.

The insertion of an electrical command signal in the dielectric rods, which function as diodes, increases the conductivity, but the optical loss is very small. On the other hand, the increase in the conductivity generates an increase in the refractive index. The tuning of the required optical channel is achieved by means of the proper voltage inserted in the diodes.

Chapter 4 - Photonic crystals (PCs) are periodically structured electromagnetic media, possessing photonic band gaps, i.e. ranges of frequency in which light can not propagate through the structure. The periodicity of the structure is of the order of the wavelength of light. The photonic band gaps are analogous to the band gaps or energy gaps for electrons traveling in semiconductors. In the case of semiconductors, a band gap arises from the wave-like nature of electrons. Electrons as waves within a semiconductor experience periodic potential from each atom and are reflected by the atoms. Under certain conditions, electrons with certain wavevectors and energy constitute standing waves. The range of energy, named "bandgap" in which electrons are not allowed to exist, appears. In a similar manner, standing waves of electromagnetic waves can be formed within a periodic structure whose period is of the order of the light wavelength. In this case, the medium expels photons with certain wavelengths and wavevectors. Such a structure acts as an "insulator of light", and this phenomenon is referred to as photonic band gap formation.

Chapter 5 - This chapter demonstrates a novel way to design high performance tunable reflectors by modifying a 1D binary PBG (Photonic Band Gap) structure to a 1D ternary periodic structure. This approach significantly enhances the incidence angle based tunability of reflection bands in the structure. This incidence angle based tunability of reflection bands can be further enhanced by increasing the thickness of the sandwiched layers in the ternary PBG structure. These 1D ternary PBG structure based high performance tunable reflectors can be substituted in place of 1D binary PBG structure based reflectors where fast and wide reflection band tuning is required.

Chapter 6 - A new multilayer structure is suggested for infrared optical filtering in temperature unstable environments. Large changes in environmental temperature in operating range of filter causes a negligible effect on output spectrum. Due to superior immunity to temperature changes, this proposed new multilayer structure is a superior substitute to traditional 1D binary PBG (Photonic Band Gap) structure where temperature stable performance is required.

In: Photonic Crystals
Editor: Donald Claus

ISBN: 978-1-62618-193-9

Chapter 1

DIAMOND PHOTONICS: TURNING DISCRETE NANO-SCALE INTERACTIONS INTO INTEGRATED QUANTUM INFORMATION PROCESSING DEVICES

Igal Bayn[1], Dirk Englund[1], Joseph Salzman[2] and Rafi Kalish[3]

[1]Department of Electrical Engineering, Columbia University, New York, NY, US

[2]Department of Electrical Engineering, Zisapel Micro-Nano-Electronics Center and Solid State Institute, Technion Haifa, Israel

[3]Department of Physics and Solid State Institute, Technion Haifa, Israel

ABSTRACT

The field of quantum information processing (QIP) takes advantage of the properties of quantum mechanics to perform tasks that are classically challenging, including exponentially faster computational algorithms, secure communication, and complex physical and biological modeling. A number of recent theoretical and experimental advances have identified promising candidates for the implementation of QIP. Although the quest for an optimal physical system to encode quantum information is far from being over, the negatively charged nitrogen-vacancy color center in diamond (NV^-) presents significant advantages over its counterparts. This is the only system, reported so far, in which a single spin can be initialized, manipulated, optically read out, all at room temperature. Moreover, the storage time can amount to thousands of gate durations (2 ms for electon spins, up to 1 second for nuclear spins - the latter would enable more than 10^3 logic gates). Indeed, several groups demonstrated single and two-qubit operations with NV^- in diamond, thus, meeting all DiVincenzo conditions for a quantum bit, besides *scalability*. In this chapter we review photon-assisted interactions between distant NVs as a way to overcome the scalability problem in diamond NV quantum processing. This implies the ability to convert spin states in NV^- stationary qubits into photon states and vice versa, combined with a reliable transfer of photons between specific NVs [not quite – let's talk about this point]. This interaction can be greatly facilitated in an integrated quantum photonic architecture (including photonic crystal cavities and waveguides) for confining photons around NV centers and connecting between distant NV sites. Here we review the advancements in the field of diamond photonics in the last decade. This includes photonic structure design, various fabrication approaches towards nano-scale processing in a single crystal diamond substrate, and the state-of-the-art experimental demonstrations of single and integrated photonic elements.

INTRODUCTION

Quantum Information Technology (QIT) may revolutionize many classical electronics areas, amongst others computing and communication. QIT offers solutions to problems that nowadays computers are inefficient in solving and may enable exponentially faster computational algorithms, communication secured by the laws of physics, and modeling of complex quantum and biological systems [1, 2]. Quantum information is generated, stored, and manipulated in two-level quantum systems, or quantum bits (qubits). Due to the inherent parallelism of quantum mechanics, quantum hardware of N qubits allows an N-qubit register to be in 2^N states simultaneously. However, despite the great appeal of the exponential computational power of quantum mechanics with the number of qubits involved, the realization of both computation and storage capabilities in many-qubit system is challenging.

One attractive approach towards realization of this task introduces a *quantum network,* as proposed by *J. I. Cirac et al* in Ref. [3]. The network is formed by stationary quantum *nodes* in which information is stored and locally manipulated. Each node is composed of one or multiple stationary qubits with long coherence time, such as those in atomic states [3].

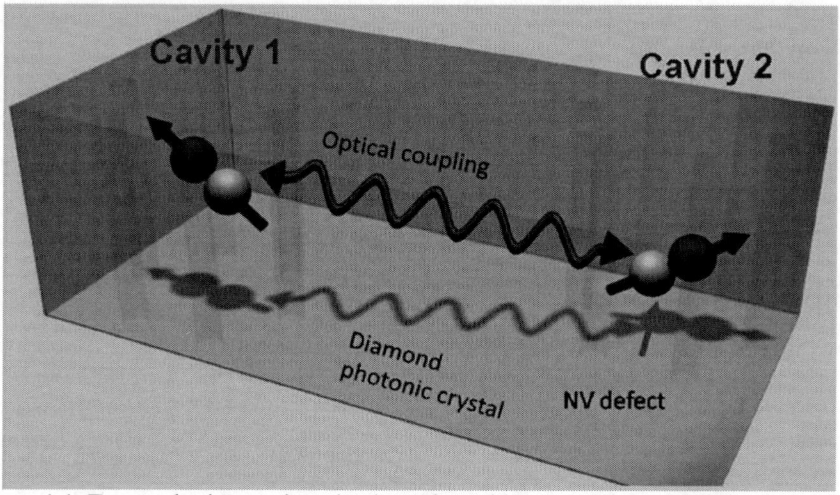

Figure 1.1. Two nodes interaction via photonic architecture. Each node is composed by a single qubit (NV^-). The qubits are strongly coupled to high-Q cavities connected through a photonic waveguide [75].

The nodes are interconnected by photonic channels that provide long-range interaction. One appealing way to do so is by registering each node into a high quality factor (*Q*) cavity where it is strongly coupled to the cavity mode. Then, reversible coherent state transfer between the photon and the node becomes possible [4]. This, potentially, allows distribution of quantum entanglement across the entire network [5].

Naturally, the number of qubits in each node will be limited and qubits may present certain non-uniformity in their energy levels and coherence times. An ideal quantum photonic interface would not only allow connecting the best-fitted qubits while bypassing the others, but would also enable increase in processing power.

For instance [5], if the network of *m* nodes with *p* qubits in each one is assumed for the quantum interface between the nodes, a state space of 2^{mp} is obtained, while for the classical one the space reduces to $m \times 2^{p}$. This distributed computation extension to the quantum network idea was introduced by *H. J. Kimble* in Ref. [5] and denoted as *quantum internet*.

The basic scheme of the quantum interconnect between two separate nodes in diamond is shown in Figure 1-1. Each node consists of one qubit, which is formed by a negatively charged nitrogen-vacancy color center (NV^-) in diamond.

The qubit is strongly coupled to the high-Q cavity and two cavities are controllably interconnected through a waveguide. The realization of this scheme all in the same material (i.e. in diamond) has potentially many advantages, as will be discussed below. A snapshot on the present development of this photonic architecture is the main goal of this chapter. In the following sections, we will give a brief introduction on quantum computing and explain how this scheme may be implemented in diamond and why this is a material of choice for QIT.

Primer into Quantum Computing (QC)

Before discussing the realization of photonic crystals in diamond, some introduction into the basic concept of QC is required. This section provides a succinct and basic flavor of QC, stressing a few essential definitions that will be needed later.

Quantum Bit (Qubit)

A qubit may be any quantum system that has at least two states [2]. These are called and $|1\rangle$. The states shall be orthogonal ($\langle 0|1\rangle = 0$). The main difference between the quantum and the classical bit is that the quantum can be initialized into a *coherent superposition*:

$$|Q\rangle = a_0|0\rangle + a_1|1\rangle = \begin{pmatrix} a_0 \\ a_1 \end{pmatrix} \tag{1.1}$$

where a_0 and a_1 are amplitudes defining the probability of the qubit to be in the states and $|1\rangle$, respectively, and satisfy the normalization condition $|a_0|^2 + |a_1|^2 = 1$.

A qubit state may by described by its amplitude vector (see Eq. 1.1). A set of N qubits forms a *quantum register* of size N.

Single and Two Qubit Operations [1]

Single Qubit Operations: The single qubit operations are described by 2×2 matrices using the vector representation. In fact, any *unitary* matrix *R*, is a valid quantum operation. The unitarity assures that all quantum operations are *reversible*, meaning that the information is neither created nor deleted and it is possible to return to the state via $R^{\dagger}$, the unitary operation ($R^{\dagger}R=1$). The number of single qubit operations is infinite, just as the space of 2×2 unitary matrices. Multiple Qubit Operations: First we introduce the controlled-Not (CNOT) gate which may be regarded as a quantum analogue of the classical XOR gate. This gate operates on two qubits: control (*A*) and target (*B*).

The control qubit is left intact, while on the target qubit the XOR operation of both bits is performed ($|A\otimes B\rangle$), i.e., it flips the target qubit *B* if the control qubit *A* is in a state $|1\rangle$.

The CNOT gate combined with single qubit operations can produce any multiple qubit gate. Therefore, it is considered as a set of universal gates. To implement complex quantum algorithms [1] one needs to realize these gates in a multi-qubit environment. For a true-quantum computer these have to be augmented with storage, initialization and read-out capability, as explained below.

Decoherence

The description of quantum computation presented above is over-simplified since qubits were assumed to be isolated entities interacting with each other without being affected by the surrounding environment.

In practice, the coupling to the environment influences quantum states, introducing mechanisms through which the information is irreversibly lost. These can be divided into two categories, each characterized by its own time scale:

- *T_1 is a lifetime of the excited state* $|1\rangle$. Through this loss process the state undergoes the transition to state $|0\rangle$, or any other state. The transition may be radiative or non-radiative.
- *T_2 is a phase coherence time.* In this case the populations of the states $|0,1\rangle$ stay intact, but their relative phase varies. Since quantum information relies on a *coherent superposition* of both states, the information is lost.

Certainly, the minimal coherence time defines an upper bound for the number of quantum operations that can be performed on a single qubit. Therefore, if the time for a single operation is given by T_{op}, the total operation number is limited by:

$$N_{op} = \frac{\min(T_1, T_2)}{T_{op}} \tag{1.2}$$

DiVincenzo Conditions

To derive practical requirements for the quantum computation hardware, a set of criteria has been introduced by D. P. DiVincenzo (see Ref. [6]):

(i) A scalable physical system with well characterized qubits.
(ii) The ability to initialize the state of the qubits to a simple fiducial state, such as $|000....\rangle$.
(iii) Long relevant decoherence times, much longer than the gate operation time (T_{op}).
(iv) A "universal" set of quantum gates.
(v) A qubit-specific measurement capability.

Two additional requirements were added to facilitate quantum communication:

(i) The ability to interconvert stationary and "flying qubits" (photons).
(ii) The ability faithfully to transmit flying qubits between specified locations.

Several possible realizations of quantum operations on various qubits have been reported. These were based on photons [7], trapped ions [8, 9], quantum dots [10], and superconducting charge devices [11]. Although photon supports many qubit-shared entanglement, which is important in QC, quantum information storage with photons is challenging. Among the solid-state alternatives, we will show that the diamond NV^- center not only satisfies most of the criteria coined above, but also introduces an opportunity for room temperature operation.

S*calability* is the most difficult requirement for qubit implementation. Surprisingly, a possible solution for this problem is based on harnessing quantum communication satisfying the (vi) and (vii) requirements above and integrating them into QC hardware.

Diamond Nitrogen Vacancy (NV) Color Centers

Unique properties of diamond such as extraordinary hardness, very high thermal conductivity, optical transparency ranging from ultra-violet to infra-red, chemical and radiation resistance, high electron and hole mobility, and its everlasting allure as a gemstone, are well known. However, the potential of diamond as a material of choice for QIT has not been seriously considered until recently. One class of the material's exceptional quantum properties originate from defects or impurities introduced into the crystal. These are called color centers, since they are responsible for brilliance in diamonds, rubies and emeralds. Among the more than 100 color centers in diamond [12], the negatively charged nitrogen vacancy color center (NV^-) is of a special importance, since it allows *spin manipulation at room temperature*, an important attribute for enabling practical QC applications. The NV center consists of a nitrogen atom substituting a carbon one adjacent to an adjacent lattice vacancy; the NV^- center is a negatively charged NV (see Figure 1-2a). This lattice point defect forms an atom-like electronic wavefunction with well-defined spin and angular momentum. The NV^- has two unpaired electrons that constitute spin=1 magnetic sub-level forming spin-riplet with ^3A and ^3E symmetries in the ground and excited states [13]. The states are denoted by total spin: $m_s=0$, ± 1.The ground state is split into ($m_s=0$) and ($m_s=\pm 1$) sublevels separated by 2.88GHz (Ref. [14]), as is shown in Figure 1-2b.

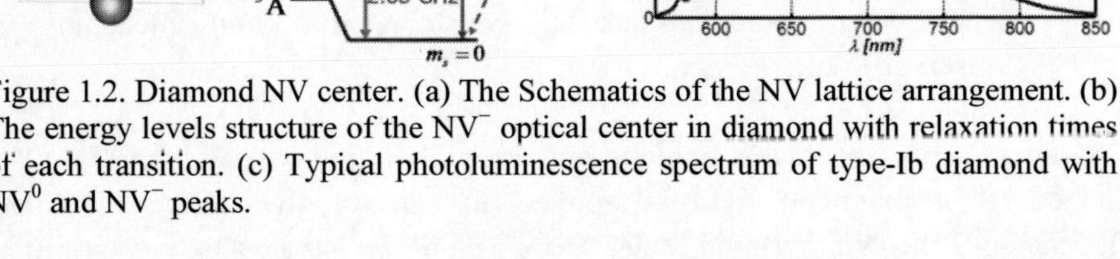

Figure 1.2. Diamond NV center. (a) The Schematics of the NV lattice arrangement. (b) The energy levels structure of the NV^- optical center in diamond with relaxation times of each transition. (c) Typical photoluminescence spectrum of type-Ib diamond with NV^0 and NV^- peaks.

The optical transition between the $m_s=0$ ground and excited states has a zero-phonon-line (ZPL) energy of 1.945eV ($\lambda_0=638nm$), which allows convenient excitation via standard red diode laser. This transition exhibits extremely high quantum efficiency, allowing single defect spectroscopy at room temperature [15]. The NV^- shows a high degree of spin polarization under optical excitation [14]. For instance, polarization into a ground state $^3A(m_s=0)$ appears due to existence of a singlet electronic state (^1A) whose energy lies between the ground and the excited triplets. Transition into the ^1A state occurs predominantly from $^3E(m_s=\pm 1)$, while it decays only into the $^3A(m_s=0)$ state [16]. This decay mechanism allows the detection of a single spin. If non resonant excitation is used (for example, green laser of 532nm), all ground triplet states are excited $^3A(m_s=0,\pm 1)$. However, only $m_s=\pm 1$ pass intersystem crossing into ^1A, thus producing fewer fluorescence cycles before shelving than its $m_s=0$ counterpart. This amounts to 20-40% decrease in photoluminescence (PL) [17]. Applying a microwave field tuned to the $2.88GHz$ transition between $^3A(m_s=0)$ and $^3A(m_s=\pm 1)$ enables spin manipulation into any superposition of upper and lower states [13].

Note: In the PL signal (Figure 2c) another peak of 575.5nm is clearly seen. This is the neutrally charged nitrogen-vacancy color center (NV^0). In general, the charge state of the NV is significantly influenced by the diamond surface [18, 19] as is described later in this chapter. For our purpose, the non charged NV^0 defect does not possess the desired properties for QIT applications.

Among other important properties of NV^- are:

- Low decoherence of ^3A states: Spin-lattice relaxation time T_1 of 10^2s and coherence time T_2 of 1.8ms at room temperature [20]. These are the longest values for any solid-state electron spin qubit. They originate from weak spin-orbit interaction and a spin free neighborhood of NV center in a highly pure single crystal diamond.
- The diamond matrix can withstand extremely high electric fields, thus allowing the use of the optical transition Stark shift for frequency tuning [21].
- The optical center is particularly stable (no photo bleaching is observed) [22].

These properties make diamond an extremely attractive candidate for QC. Indeed, if a magnetic field is applied to remove the $^3A(m_s=\pm 1)$ state degeneracy, the NV^- ground states (for instance $m_s=0$ *and* $m_s=-1$) form a two-level system used to encode quantum information (qubit). This can be read-out optically, initialized and manipulated, all *at room temperature*. Furthermore, long coherence time makes performance of over 10^6 single-qubit operations possible [23]. These properties allowed recent demonstration of Rabi oscillations [24] and simple two qubit operations [24, 25], generating increasing large quantum registers [26-28] and NV-photon coupling [29]. As a matter of fact, diamond NV^- satisfies all DiVincenzo criteria, besides *scalability*.

DIAMOND QUANTUM COMPUTING PHOTONIC ARCHITECTURE

Although Rabi oscillations [24], single- / two-qubit operations [24, 25] and quantum registers [26-28] with NV^- were demonstrated, the realization of large scale quantum computing systems requires a controllable interaction between distant qubits which cannot rely on the dipole-dipole mechanism (restricted by a ~*10* nm maximal distance between the centers), as was used in the above experiments. Furthermore, most of these demonstrations were based on a random sample search for an appropriate configuration of the NVs and adjacent nuclei. This approach cannot be extended into a scalable many qubit integrated QC. That is why the interaction between distant qubits in a multi-qubit operation should be carried out by photons using the quantum network approach [3-5]. To implement the idealized two-node-network from Figure 1-1, one may opt for a photonic crystal (PC) architecture (see Figure 2-1), due to the inherent scalability and convenience in the network element design. For instance, PC cavities and waveguides [30, 31] can be produced by omitting or shifting several holes in the band-gap region (see the discussion in Section 4). Moreover, such a system is an ideal building block for a photonic module [32], which provides a natural method for generating a three-dimensional cluster state based quantum computer architecture [33].

In Figure 2-1 each node is composed of a single NV^- qubit coupled to a high-Q cavity in the Purcell regime, in which the excited state predominantly couples to one optical mode [30, 31]. The node information is manipulated by applying a microwave pulse (~2.88GHz) resonant to the $^3A(m_s=0)\leftrightarrow {}^3A(m_s=-1)$ transitions (see Figure 1-2b, certain static magnetic field is applied to split $^3A(m_s=\pm 1)$). The qubit initialization to the ground state is produced by broad-band excitation [14].

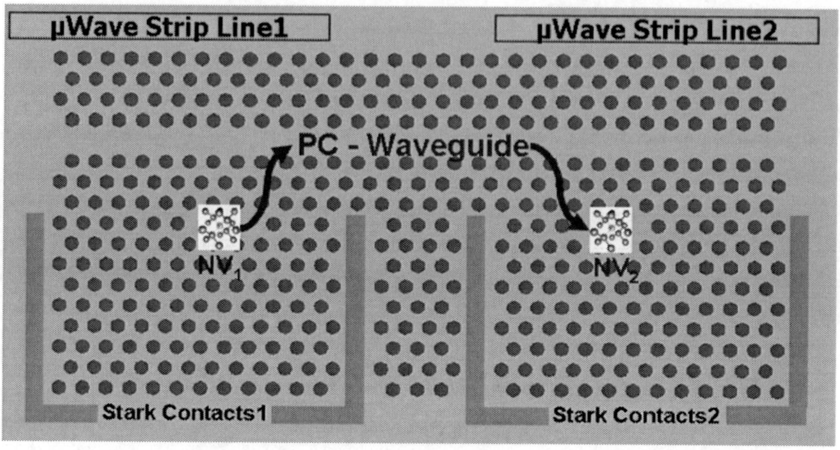

Figure 2.1 [75]. Schematics of two Qubits interconnected by a PC waveguide and manipulated via micro-wave and Stark shift controls [30, 31].

The waveguide-cavity coupling is governed by tuning the cavity resonance using the Stark effect [21, 30, 31]. The cavity coupling has two primary benefits: (i) the ZPL branching ratio can be enhanced dramatically (far beyond the 3% of the uncoupled NV^-), allowing distant NV^- *coherent* photon-mediated interactions; (ii) the spontaneous emission coupling efficiency to the cavity mode, $\beta=\Gamma_c/(\Gamma_c+\Gamma_{PC})$, approaches unity for a large F_{cav}, meaning that nearly all emission is collected into the cavity mode. Here, F_c is maximum cavity Purcell factor defined by a cavity quality factor and mode volume), $F_{PC}=\Gamma_{PC}/\Gamma_0<1$ accounts for PC spontaneous emission suppression and Γ_c/Γ_{PC} are NV^- spontaneous emission rates into a cavity and PC modes (see Section 3.1 for further discussion).

In this scheme, the requirements for large-scale entanglement, single qubit readout and quantum data storage are met. In addition, the interaction distance limited by the waveguide losses can in principle reach hundreds of microns or even several kilometers if coupled into optical fiber. Furthermore, the reversible coherent state transfer between the nodes [4] and two qubit CNOT operation [34], based on a Λ-type system where one transition of the Λ is coupled to the cavity mode and another is weakly excited, are applicable. These operations are based on the cavity-cavity or cavity-waveguide-cavity coupling. Combining them with quantum register design [30] and multi-cavity architecture interconnected by the waveguide network produces a core operation packet for quantum computer realization. This realization shares a similar approach to microelectronics, which is the most successful experience in large-scale integration to date. In addition to the long-term goal of quantum computing, the implementation of a cavity coupled to a single NV^- center in diamond can demonstrate a variety of interesting phenomena, such as a basic study of cavity quantum-electro-dynamics (QED) [35], cavity enhanced single spin non-demolition read out [36], a range of medical applications related to nanomagnetometry [37, 38] (as a heart pace maker or brain activity monitor), and bio-sensing probes of living cells [39].

DIAMOND PHOTONICS – INGREDIENTS

In this Section, the essential ingredients involved in the realization of a cavity-NV^- coupling are discussed. Included is a brief description of atom-photon interaction in the cavity, followed by the review of two main approaches towards the formation of high-Q cavities in diamond: photonic crystals and nanobeams.

Cavity QED

The interaction mechanism between the NV^- and a cavity mode can be divided into two regimes of strong and weak coupling between the emitter and the cavity mode. For efficient optical interactions between photon qubits and the NV spin state, the Purcell regime of emitter-cavity coupling is necessary. In this regime, the Purcell factor F>>1 so that the emitter interacts primarily with one optical mode, i.e., the cavity mode. The Purcell regime can be achieved even when the cavity loss rate exceeds the emitter-cavity coupling rate. This weak-coupling regime can be attained more easily than the strong coupling limit (where the emitter-coupling rate must exceed the cavity loss) and will be considered here. Furthermore, weak coupling implementation allows maturing of currently fabrication technology, and we regard the demonstration of the weak coupling as an initial step towards strong coupling. Here, both regimes are described, and their boundary condition is presented.

Weak Coupling Regime (Purcell Effect)

The rate of spontaneous emission in the cavity in the weak coupling regime is proportional to the local optical mode density of states [40]. A PC architecture can significantly modify this density [41]. In the slab configuration (see Section 3.2), such modification can significantly decrease spontaneous emission rate [42, 43]. In contrast, in a PC cavity coupled to an atom, the spontaneous emission rate is increased, while the decoupled emission rate is decreased [42]. If the spontaneous emission rate in the cavity is denoted as Γ, while the emission rate in the bulk is Γ_0, the spontaneous emission enhancement is given by [42]:

$$\frac{\Gamma}{\Gamma_0} = F_{cav}\left(\frac{\vec{E}(\vec{r}_{NV})\cdot\vec{\mu}}{\left|\vec{E}(\vec{r}_{NV})\right|\left|\vec{\mu}\right|}\right)^2 \frac{1}{1+4Q^2\left(\frac{\lambda_{NV}}{\lambda_0}-1\right)^2} + F_{PC} \tag{3.1}$$

where, $\vec{\mu}$ is the NV^- center dipole moment, which is located in $\vec{r}_{NV}$; is the electric field at the NV^- location; is NV^- wavelength (*~638nm*); and λ_0 is the cavity mode wavelength ($\lambda_0 \neq \lambda_{NV}$ due to the fabrication processing).

In the case of polarization and wavelength matching between the cavity and the NV,

$$\frac{\Gamma}{\Gamma_0} = F_{cav} + F_{PC} \tag{3.2}$$

where the F_{PC} is Purcell factor of the PC slab given by $F_{PC} = \Gamma_{PC}/\Gamma_0$ is PC slab spontaneous emission rate) and F_{cav} is Purcell cavity factor given by:

$$F_{cav} = \frac{3}{4\pi^2}\frac{\lambda_0^3}{n^3}\frac{Q}{V_m} \tag{3.3}$$

where n is slab refractive index and V_m is cavity-mode-volume given by:

$$V_m = \frac{\int_V \varepsilon(\vec{r})\left|\vec{E}(\vec{r})\right|^2 d^3\vec{r}}{\max\left[\varepsilon(\vec{r})\left|\vec{E}(\vec{r})\right|^2\right]} \tag{3.4}$$

For example, for diamond with $n=2.4$, $Q=10^6$, and $V_m=(\lambda/n)^3$, a value of $F_{cav}\approx 7.6\times 10^4$ is obtained, and thus, a large spontaneous emission enhancement is possible.

Strong Coupling Regime

Let us present a single NV center transition ($^3E\rightarrow^3A$) coupled to a high-Q cavity mode as a two level system. The cavity mode (ω_0) and the two-level system transition ($\hbar\omega_{NV} = E_b - E_a$) are in close proximity, so that $\omega_{NV} \approx \omega_0$. If non-resonant processes are neglected (processes with a gain or loss of energy) and constant energy terms are disregarded, the system is described by the Jaynes-Cummings Hamiltonian [35]:

$$\hat{H}_{JC} = \hbar\omega_{NV}|b\rangle\langle b| + \hbar\omega_0\hat{a}_l^\dagger\hat{a}_l + i\tfrac{1}{2}\hbar\Omega_v\left(|a\rangle\langle b|\hat{a}_l^\dagger - |b\rangle\langle a|\hat{a}_l\right) \tag{3.5}$$

where: $|b\rangle$, $|a\rangle$ are the atom energy levels with the corresponding frequencies E_b, $(E_a=0)$ and $a_l^\dagger$, a_l are the creation and annihilation operators of the cavity mode ω_0. Ω_v is the vacuum Rabi frequency given by $\Omega_v \equiv 2E_l d_{ab}/\hbar$ and is the dipole moment operator ($d_{ab} = \langle a|\hat{D}_z|b\rangle$, assuming that the cavity mode is polarized in $\hat{z}$).

The eigenstates of this Hamiltonian are "dressed states." Assuming that n_p is a photon number, $\hat{H}_{JC}$ is diagonalized in the groups: $M_n = \left\{|a,n_p\rangle, |b,n_p-1\rangle\right\}$. Therefore,

$$\begin{aligned}
&\hat{H}_{JC}\left|\psi_{\pm n_p}\right\rangle = E_{\pm n_p}\left|\psi_{\pm n_p}\right\rangle \\
&E_{\pm n_p} = \hbar\left(n_p\omega_l - \tfrac{1}{2}\delta \pm \tfrac{1}{2}\sqrt{n_p\Omega_v^2 + \delta^2}\right)
\end{aligned} \tag{3.6a,b}$$

$$\begin{cases}
\left|\psi_{+n_p}\right\rangle = \cos\theta_{n_p}\left|a,n_p\right\rangle + i\sin\theta_{n_p}\left|b,n_p-1\right\rangle \\
\left|\psi_{-n_p}\right\rangle = i\sin\theta n_p\left|a,n_p\right\rangle + \cos\theta_{n_p}\left|b,n_p-1\right\rangle
\end{cases} \tag{3.7}$$

where: $\delta' \equiv \omega_l - \omega_{NV}$, $\tan 2\theta_{n_p} \equiv -\frac{\Omega_v\sqrt{n_p}}{\delta}$.

In contrast to the semi-classical treatment [40] where radiation modulates only the atomic transition, here the eigenstate describes both the radiation and the (atomic) two level system. These are "dressed states", meaning that

$\left|\psi_{\pm n_p}\right\rangle \neq \left|\psi_{\mathrm{NV}}\right\rangle \otimes \left|\psi_{\mathrm{rad}}\right\rangle$, and the energy splitting (Eq. 3.6b) can be clearly detected optically using weak probing [44].

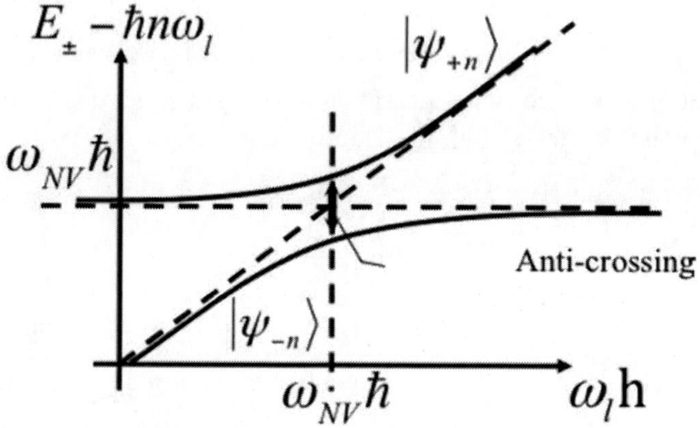

Figure 3.1. Strong-Coupling-Regime: Eigen energy diagram.

Certainly when the detuning is increased, the dressed states become more photon-like or atom-like (see Figure 3-1). To examine the dynamic behavior in the strong-coupling, we assume that $\theta_{n_p} = \frac{\pi}{4}$, $\delta' = 0$, which results in:

$$\begin{aligned}\left|\psi_{+n_p}\right\rangle &= \tfrac{1}{\sqrt{2}}\left(\left|a,n_p\right\rangle + i\left|b,n_p-1\right\rangle\right)\\ \left|\psi_{-n_p}\right\rangle &= \tfrac{1}{\sqrt{2}}\left(i\left|a,n_p\right\rangle + \left|b,n_p-1\right\rangle\right)\end{aligned} \tag{3.7}$$

Thus, for the initial condition: $\left|\psi(0)\right\rangle = \left|b,0\right\rangle$, we obtain: $\left|\psi(0)\right\rangle = \frac{1}{i\sqrt{2}}\left(\left|\psi_{+1}\right\rangle + i\left|\psi_{-1}\right\rangle\right)$ which provides the temporal behavior

$$\left|\psi(t)\right\rangle = e^{i\omega_{NV}t}\left(\cos\tfrac{\Omega_1 t}{2}\left|b,0\right\rangle - \sin\tfrac{\Omega_1 t}{2}\left|a,1\right\rangle\right) \tag{3.8}$$

Equation (3.8) describes Rabi oscillations. In contrast to the semi-classical treatment [40], the oscillations are between dressed states and not between atomic states. Moreover, even zero photon radiation ($n_p=0$) can produce the oscillations.

QED Conditions

The strong coupling regime has been demonstrated for a variety of cavity designs [42, 45-50]. Here we evaluate the boundary conditions of this regime for the diamond NV^{-} coupling. Radiation mode losses originate from the finite photon life-time in the cavity, which results in the finite cavity quality factor (Q). The cavity mode decay rate κ is given by $\omega_0/4\pi Q$. The atomic losses originate from the coupling of the atom to modes other than the cavity one. This atom decay rate is given by $\gamma_{\perp}$, which is the NV^{-} dipole decay rate. A rigorous treatment of this loss mechanism is given elsewhere [35, 40]. The condition for the strong coupling regime is $\Omega_r >> \kappa, \gamma_{\perp}$; when this condition is not satisfied, the NV center is in the weak-coupling regime. The Ω_r and $\gamma_{\perp}$ are related via $\Omega_r = \gamma_{\perp}(V_0/V_m)^{1/2}$, where $V_0 = c\lambda^2/(8\pi\gamma_{\perp})$ [51]. Adopting $\gamma_{\perp} = 32MHz$, [30] and following the argumentation of Ref. [52], we obtain that strong coupling is achieved when $Q/V_m^{1/2} >> 2.5\times 10^3$.

Diamond Photonic Designs

Photonic Crystals

The concept of photonic crystals (PC) was first introduced by Eli Yablonovitch and Sajeev John in their seminal works [41, 53]. The PC is formed by a periodic modulation of the dielectric constant [54]. The solution of the wave-equation for PC presents frequency bands and gaps between them [55]. These gap regions are called photonic band-gaps (PBG). In the PBG frequencies, light propagation is prohibited and the photonic density of states is reduced [54]. As a result, the spontaneous emission rate in the PBG frequencies is substantially decreased due to the Purcell effect [41]. Only three-dimensional (3D) PCs exhibit full PBG, prohibiting propagation in all directions for both transverse-electric (TE) and transverse-magnetic (TM) polarizations [41, 53, 54]. Realization of 3D-PC is extremely challenging [56, 57]. Therefore, slab based devices are considered in this work.

A slab PC is formed by a membrane suspended in air and modulated periodically by a triangular array of cylindrical air holes. In this structure the bands can no longer be designated as pure TE or TM. Instead, they form two groups: TE-like (*y-even*) and TM-like (*y-odd*), in accordance to their parity in the mid-depth plane (*y=0,* see Figure 3-2a) [51, 58]. The light-confinement is produced by PC and total-internal reflection (TIR). The modes of the slab can be divided into guided and radiative. The guided modes are localized in the membrane by TIR, while the radiative ones leave the slab in the *y* direction.

In a PC, the in-plane wave-vector $\vec{k}_l = \left(k_x, k_z\right)$ is defined by the inverse-lattice [54] and the dispersion curve is fully described within the first-irreducible-Brillouin zone, forming a triangle with high-symmetry point vertices (*Γ,X,J*) (see Figure 3-2b). The light cone separating between the radiative and confined modes is given by $\omega_{LC}\left(k_x, k_z\right) = c\sqrt{k_x^2 + k_z^2}$. As one can observe from Figure 3-2c, the maximum PBG is obtained between the 1st and the 2nd TE-like modes in diamond [59, 60]. Thus, a PC cavity resonant frequency is preferentially designed within this gap.

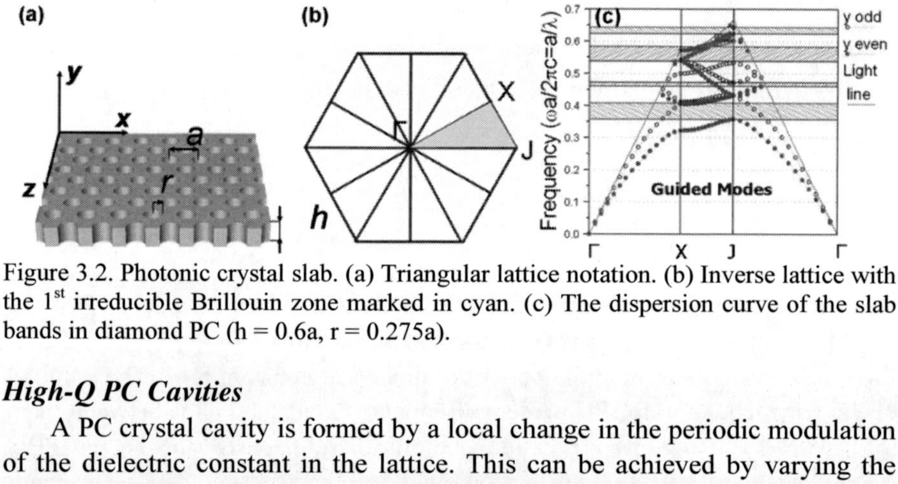

Figure 3.2. Photonic crystal slab. (a) Triangular lattice notation. (b) Inverse lattice with the 1st irreducible Brillouin zone marked in cyan. (c) The dispersion curve of the slab bands in diamond PC (h = 0.6a, r = 0.275a).

High-Q PC Cavities

A PC crystal cavity is formed by a local change in the periodic modulation of the dielectric constant in the lattice. This can be achieved by varying the hole position, size or form. The process is similar to the introduction of an atomic defect in a semiconductor. Note that by producing a line of defects, a photonic waveguide is formed [61]. The high-*Q* cavity design is more "delicate", thus demanding a deeper insight into the light confinement. In this section a very succinct introduction of high-*Q* cavity design is given. The cavity quality factor can be divided into two major components: Q_l (Q_v) for the lateral (vertical) losses. [51, 62] Q_l is defined by the in-plane losses (*x-z*) and determined by the number of PC periods around the cavity. The increase in the number of periods leads to an unbound increase in Q_l. Thus, by embedding the cavity in a sufficiently large PC, the value of Q_l can be made arbitrarily large, and the total quality factor becomes limited by Q_v only [51, 62], i.e. $Q \approx Q_v$.

The vertical losses are defined by the power radiated in the y direction. Since we focus here on the TE-like mode confined in the first-order PBG, it can be shown that the vertically radiated power loss that defines the Q_v follows from the plane-wave components of the cavity mode that are within the light cone and thus not confined by TIR [51, 62]:

$$P_v \approx \frac{\eta}{2\lambda_0^2 k}\int_{k_l \le k} \frac{dk_x dk_z}{k_l^2} k_z \left|FT_2\left(H_y\right)\right|^2 \tag{3.9}$$

where: $\eta=(\mu/\varepsilon)^{1/2}$, λ_0 is the cavity mode wavelength, $k=2\pi/\lambda_0= [k_l^2+k_y^2]$, $k_l=(k_x\ k_z)=k(sin\theta cos\varphi\ sin\theta sin\varphi)$ and $k_y=kcos\theta$. H_y is the y-component of magnetic field above or below the membrane ($y=\pm h/2$). Note that the circle $k_l<k$ defines the light cone edge, thus separating between the localized and radiative components of the mode. In general, due to the uncertainty principle, any mode localization in real space (forming a cavity) produces mode delocalization in k-space. Thus, one may treat cavity Q optimization as an attempt to minimize the overlap of the $|FT_2(H_y)|$ distribution with the light cone, thus decreasing the vertical losses. This process is schematically shown in Figure 3-3.

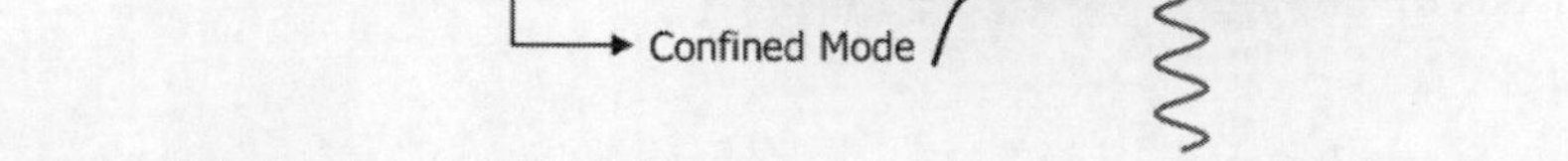

Figure 3.3. Schematics of Q optimization in the cavity by shifting the mode distribution ($|FT_2(H_y)|$) from light cone region (green) into the guided mode region (blue). On the right, a PC waveguide in Γ-J direction with radiative and confined modes is shown.

To date, the PC cavity designs with highest Q/V_m ratio are based on an optimized spatial variation of the dielectric constant in the vicinity of PC-waveguides (see Figure 3-3). This design was first introduced by B. Song *et al* in Ref. [63], displaying a numerical value of $Q=2.4\times10^7$ with $V_m=1.3\times(\lambda/n)^3$ in silicon. In later works, several variations on this technique were developed

[62, 64]. The state of the art silicon fabrication [65] exhibits $Q=2.5\times10^6$, which is ~10 times lower than the modeled value, due to fabrication imperfections [66].

Diamond PC Designs

One of the very first suggestions to register the interacting diamond NV^- in one big cavity for QC was introduced by M. S. Shahriar *et al* in 2002 (see Ref. [34]). However, the first introduction of photonic architecture as a possible solution for scalability was suggested by Salzman *et al* in 2006 in Ref. [30] where a photonic crystal (PC) implementation was proposed. At this stage, it was not clear whether the diamond's moderate refractive index ($n=2.4$) could allow for *high-Q* cavities required to realize the strong coupling cavity QED regime [30] and whether this design could be fabricated. The answer the first qustion relies on heavy computational effort that will be reviewed later in this section; however, prior to this review we present a very succinct introduction to the computational methodology.

Photonic Modeling

Two of the common computational methods used to model electromagnetic modes in PCs and nanobeams are the plane-wave expansion (PWE) and Finite-Difference-Time-Domain (FDTD) simulations.

Plane Wave Expansion Method (PWE)

According to Bloch theorem [54], magnetic field in periodic media is given by $\vec{H}(\vec{r})=e^{i\vec{k}\cdot\vec{r}}\vec{u}_k(\vec{r})$, where $\vec{u}_k(\vec{r})$ is a periodic Bloch function, with the same periodicity as the lattice. Substitution of this into the master equation [54] leads to:

$$\hat{L}\vec{u}_k(\vec{r})=(i\vec{k}+\nabla)\times\left(\frac{1}{\varepsilon(\vec{r})}(i\vec{k}+\nabla)\right)\times\vec{u}_k(\vec{r})=\bar{\omega}^2\vec{u}_k(\vec{r}) \tag{4.1}$$

Where $\varepsilon(\vec{r})$ is the high-index material's dielectric constant and $\bar{\omega}=\frac{\omega}{c}$ is a normalized frequency.

This equation is solved by the PWE software in the irreducible 1st Brillouin zone in the lattice *k*-space (see for example [67]). In the case of triangular 2D PC with a lattice constant *a* and periodic hole structure in the *xz*-

plane, the zone forms a triangle with the following vertices: $\Gamma=(k_x\ k_z)=(0\ 0)$, $X=\left(0 \quad 2\pi/\sqrt{3}a\right)$, $J=\left(\pi/a \quad 2\pi/\sqrt{3}a\right)$ (see Figure 3.2 above). For a PC waveguide in *x* direction or nanobeam in *z* direction with the same periodicity, the 1st irreducible Brillouin zone is defined by $k_{x,z}=\left[0 \quad \pi/a\right]$, respectively.

Often, PCs as periodic structures are solved by the PWE method, searching the eigenvalues (frequencies) and eigenvectors (fields) of Eq. 4.1 in the first Brillouin zone in the *k*-space. However, since the slab is not periodic in *y*-direction, it cannot be calculated via this scheme without further assumptions. To permit PC slab modeling, artificial 3D periodicity is imposed by replicating the slab and its air cladding in the y direction with sufficiently large spacing between them (h_{SC}). (Typically, h_{SC} of *6-8a* is sufficient) [58]. The slab guided modes are confined in the vicinity of the membrane; thus, this artificial periodicity does not affect them significantly. The radiation modes are influenced by this new slab-lattice, but they are effectively filtered by the light cone line.

The waveguide mode calculation is very similar to the slab one, with one exception: since the waveguide in the PC is formed by modifying the refractive index or dimensions of the PC holes along a certain (waveguide) direction, an additional super-cell in the plane (*xz*) direction perpendicular to the waveguide is formed [61]. The waveguide modes are sufficiently confined to the defect line so that they are not affected by the introduction of this new super-cell.

Finite-Difference-Time-Domain (FDTD)

The FDTD method was proposed by Yee [68] and can be used to solve Maxwell's equations numerically. FDTD is a general technique, but can require large memory and computation time. Here, we first introduce the calculation of cavity *Q* and then discuss FDTD cavity calculations.

Cavity Quality Factor Q

In a resonant cavity, the ability to build up large field intensities with moderate power inputs is characterized by its quality factor *Q*. It is defined as 2π times the ratio of the time-averaged energy stored in the cavity to the energy loss per cycle, and is given by [69]:

$$Q=\omega_0 \frac{Stored\ Energy}{Power\ Loss} \tag{4.2}$$

where ω_0 is the resonant frequency.

FDTD Cavity Calculations

The PC cavity spectrum, resonant modes and quality factor calculations are all performed with a 3D FDTD method (though symmetries or approximations can be exploited to reduce the dimensionality). The PC cavity calculation is composed of three main steps:

1. Spectral Response Calculation: An impulse excitation field is applied at *t=0*. The field is evolved in time in the computation domain without any further driving, thus allowing the modes to decay. During the run, the field is stored (often only at a few low-symmetry points, to reduce memory requirements) to obtain the temporal response of the cavity. Then, by applying Fourier transform, the electromagnetic spectrum is obtained, which can then reveal cavity modes.
2. Mode Calculation: Several mode calculations with increasing computation time are performed one after another to hone into the desired cavity resonance frequency. Advanced harmonic-analysis can be applied to improve upon the Q factor estimation from a simple Fourier transform [70]. The conversion of Q is validated. The mode is filtered and the electric and magnetic fields are stored.
3. Cavity Field characteristics: The cavity parameters of interest such as the mode volume (V_m), field profiles or partial quality factors (Q_l and Q_v), are derived.

In the 3D FDTD technique, special attention is given to minimizing the error of the final result. These requires conversion study that can be divided into the following categories:

- *Spatial Resolution:* Spatial discretization is sufficient to resolve the smallest feature. Then it is decreased, paying special attention to the "sensitive" directions (*x,z* for PC and *z* for nanobeam), until the result conversion is achieved.
- *Computational Domain:* The computation spatial domain is surrounded by the absorbing boundary walls that eliminate any outward propagating energy impinging on them. These are implemented as perfectly matched layer (PML) [71]. Over this context, computation convergence is checked separately on PML

characteristic parameters (width and reflectivity) and domain dimensions.

- *Symmetry Conditions:* The usage of symmetry conditions instead of PML where possible is highly recommended since it may potentially reduce computation domain by a factor of 1/8.

Diamond PC High-Q Cavities

The initail work on diamond PCs focused on an attempt to "translate" the developed silicon cavity designs by adjusting their parameters to diamond. In this way, single point defects [59, 60, 72], three- [59] and seven-missing hole [73] and the PC waveguides and double heterostructure designs [60] were analyzed for diamond by extensive search of the numerical space parameters. This was done to optimize the Q and reduce the V_m. The analysis showed that the optimal thickness for the diamond membranes is approximately *30%* higher than for silicon counterparts and is in the region of $0.9a$-$1a$, where a is the PC lattice constant, i.e. the distance between the PC holes (see Figure 3-2a) [52]. Furthermore, for high-Q the hole radius shall be in the region of $r{\sim}0.27$-$0.3a$. As in the case of silicon, the lowest optical losses have so far been obtained in the waveguide-based cavities [74], i.e., in cavities produced via local modulation of the dielectric constant in the vicinity of PC waveguide. However, we found that a simple translation of the silicon design to diamond's refractive index of 2.4 yielded a decrease in Q by more than ~20 times (from ultra-high-Q cavities in silicon ($Q{\sim}10^6$) to moderate-Q in diamond ($Q{\sim}10^4$) [60]).

In Ref. [52] the influence of the dielectric constant of the membrane on the quality factor (Q) in slab photonic crystal ultra-high-Q cavities was examined, presenting a semi-analytical model of minimal vertical losses in the form of an ideal Gaussian field distribution in the k-space for the waveguide-based cavities. The higher value of Q in the high-ε cavity was attributed to the lower mode frequency, which results in a decreased light cone. The Q ratio in a high-ε (silicon) vs. low-ε (diamond) slab was examined as a function of mode volume (V_m). The mode volume compensation technique, where cavity mode delocalization in real space compensates the increased light cone by a better localization in k-space, was suggested showing that a moderate increase in V_m by the factor of ~1.6 in diamond produces the same losses as in the analogue design in silicon. With the help of this semi-analytical model, several types of cavities were designed. In a double heterostructure design, a $Q{\approx}2.6\times10^5$ was obtained, while the highest $Q{\approx}1.3\times10^6$ with $V_m{=}1.77\times(\lambda/n)^3$ was derived for the local width modulation cavity (see Figure 3-4) [52].

The realization of an ultra-high-*Q* design, such as the one shown above, imposes extremely strict requirements on the fabrication process in terms of hole circularity, uniformity, verticalit, and the separation of the slab from the substrate. In Table 2, we analyze the modified L3 cavity with a calculated quality factor of $Q=8,370$ and a mode volume of $V_m=0.921\times(\lambda/n)^3$ [75]. Listed are the variations of the values quoted to produce a 20% reduction in *Q*. The most stringent requirements are imposed on the departure from verticality of the walls and on variations in the radius of the PC holes. Note that these parameters describe a moderate*Q* cavity, while for the ultra-high-*Q* device, the deviations allowed are significantly lower. Still it can give a reader a qualitative understanding of the fabrication tolerances required to produce such cavities.

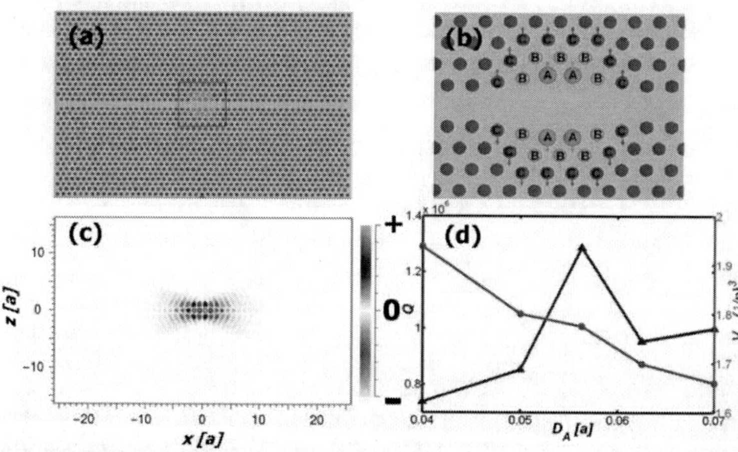

Figure 3.4. Local width modulation cavities (A1) [52]: (a) A1 cavity computational domain. (b) Detailed geometry. The holes A, B, C are shifted by D_A, D_B, D_C. (c) H_y in the plane y=0 (d) Q, V_m vs. D_A. The detailed holes shifts are summarized in Table 1.

Table 1. Local width modulation summary for the type A1 cavities in *a* units [52]

D_A	D_B	D_C
0.056	0.031	0.019

Table 2. Geometrical changes producing 20% deterioration in the cavity *Q* [75]

Design Parameter	Value for $0.8\times Q$
Wall Angle θ	1.95°
Hole radius Δr	*5nm*
Hole pitch Δa	*18nm*
Membrane Thickness Δh	*35nm*

Nanobeam Cavity and Waveguide Design

Another alternative for ultra-high-*Q* cavity design is the nanobeam geometry. The beam is based on a waveguide-like structure along the z-axis. The waveguide is physically separated from the substrate (thus forming a "bridge") referred below as a "nanobeam". A grating along the z-direction consists of a periodic set of trenches perforating the beam with periodicity *a*. This periodic structure forms a waveguide and can be regarded as a one-dimensional PC. The mode is confined to this crystal by TIR in the *xy*-plane, while its confinement/propagation in the z-direction is governed by the grating. By introducing a local variation into the grating periodicity, one produces a cavity confined in the *z*-direction by the waveguides, see Figure 3-4. Nanobeams have drawn much attention in the last years and have been implemented in various material systems, such as silicon [77, 78], gallium-arsenide [79], silicon nitride [80, 81], silicon dioxide [82] and diamond [83].

Most of these nanobeam designs required a pre-existing membrane, thus producing a rectangular cross-section beam (See Figure 3-4a). In the case of diamond, a value of $Q=3.6\times10^6$ at a mode volume of $V_m = 0.45\times(\lambda/n)^3$ has been simulated in Ref. [83].

The cavity is produced by gradually increasing the lattice constant from the center circular holes towards the surrounding waveguide claddings. This approach relies on a pre-existing diamond membrane which still poses significant fabrication challenges as discussed in the subsequent sections.

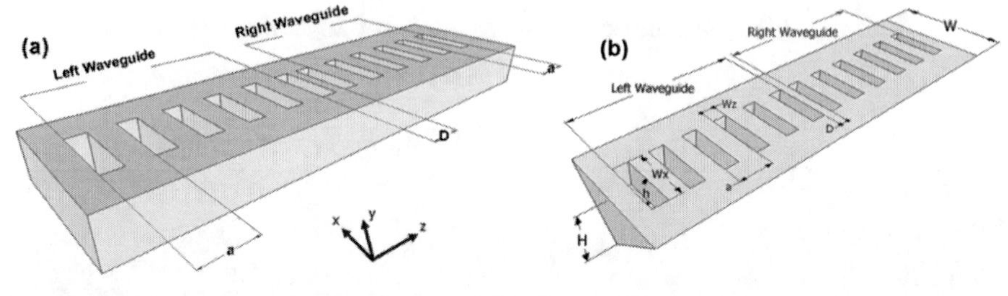

Figure 3.4. Nanobeam cavity. (a) Rectangular nanobeam with rectangular gratings forming two waveguides on the right and the left of the defect (D). (b) Triangular nanobeam [76].

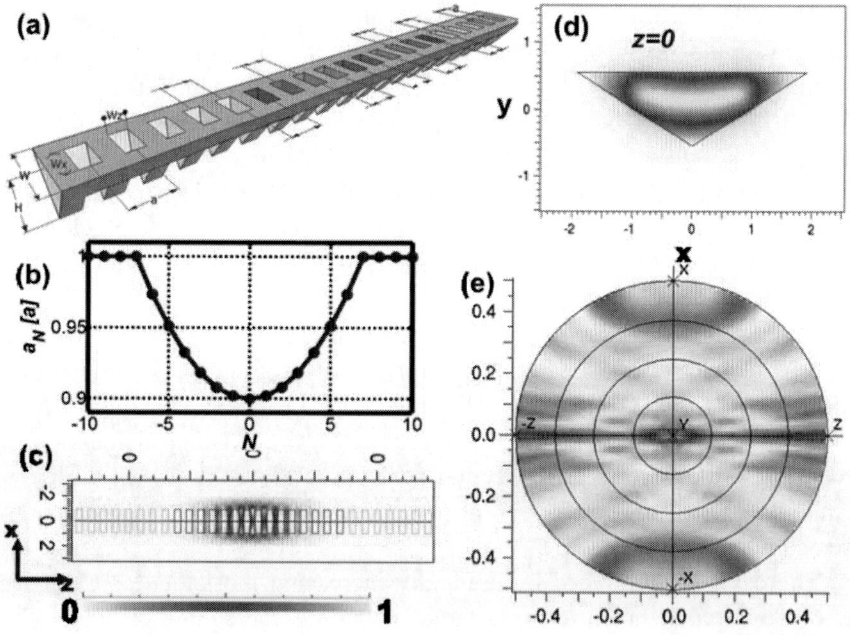

Figure 3.5. Triangular nanobeam design [84]. (a) Schematics of the triangular cross section nanobeam geometry and notation. a_0 denotes lattice constant in the defect center (a_0=0.9a), while a_N denotes gradually increased lattice constant ($a_N \leq a$). (b) The parabolic lattice constant variation responsible for the defect formation is presented ($a_N = 0.9a + (N/7)^2 \times 0.1a$. (c,d,e) Calculated cavity mode profiles of the normalized electric field energy density $U_E = \varepsilon/2|E|^2$. (c) U_E in upper y planes of the beam. (d) U_E in the z=0 plane. (e) Upper hemisphere far field distribution. The maximum appears at the vertical direction along the y axis (θ=0). All dimensions are in a units.

To bypass the difficulty in producing a diamond membrane, a design based on a triangular cross-section nanobeam [76, 84] shown schematically in Figure 3-4b was introduced [84]. This can be fabricated from a bulk substrate by inclined milling [76] or etching [85], thus eliminating the need for the membrane.

The cavity is formed by introducing a defect in the periodic structure (departure from periodicity along z) in a double hetero-structure approach, similar to that used for high Q cavities in planar PCs [63]. By decreasing the lattice constant in the defect region, the defect-mode eigen-frequency is raised with respect to that of the waveguide.

This frequency separation provides mode confinement in the z direction. In order to increase localization in k_z, the mode along z was delocalized. For that, a gentle lattice variation was designed. The lattice constant in the defect region was increased parabolically from $a_0=0.9a$ in the defect center towards a in the waveguides regions, according to the formula: $a_N=0.9a+(N/7)^2\times0.1a$ (*for* $N\leq7$) as suggested in Ref. [82] (see Fig. 3-5). This design not only produce the ultra-high-Q cavity with $Q=2.51\times10^6$, and a low mode volume of $V_m=1.062\times(\lambda/n)^3$, but also allows for preferential upward radiation, i.e. the power radiated upward into detection system objective may be $5\div20$ higher than the power radiating downward (the precise ratio depends on the design). This preferential upward radiation is extremely important for improving the potential NV-state single-shot read-out, as discussed in Ref. [76].

FABRICATION PROCESS

The best quantum properties of NVs have been reported in a single crystal diamond [86] that potentially exhibits a substantial advantage over poly-crystalline material in terms of optical losses [73]. Single crystal diamond nanophotonics faces several very substantial challenges that have prevented device realization up until recently. These include:

1. Extreme chemical inertness (no wet chemical diamond etchant).
2. Under-developed nano-patterning particularly complicated by superb dielectric and material hardness properties combined with relatively small size of the available diamond samples (2×2-5×5mm)
3. Single crystal diamond layers mostly grow on a single crystal diamond, which complicates membrane formation.

Addressing these challenges is vital in nano-photonics realization, since to confine light either in a cavity or guide in a waveguide one has to control light localization in *3D*. Such control can be achieved by PC reflection laterally and total-internal-reflection (TIR) vertically. For PCs a precise nanopatterning is required, while for TIR membrane formation processing is needed. Only realization of both of these elements leads to a successful nanofabrication.

We will review different approaches for nanopatterning with Focused-Ion-Beam (FIB) milling and electron-beam (e-beam) lithography followed by dry etching. As for the membrane formation, an ion-implantation technique, thinning of the polished diamond membranes and growth of single-crystal diamond on silicon (001) substrates with a buffer layers for PCs formation are discussed. In addition, nanobeam FIB or RIE based vertical or angular etchings are reviewed.

Patterning

In this section, we describe two patterning techniques that were developed for single crystal diamond: FIB milling and dry etching combined with photo- and e-beam lithography.

FIB-Patterning

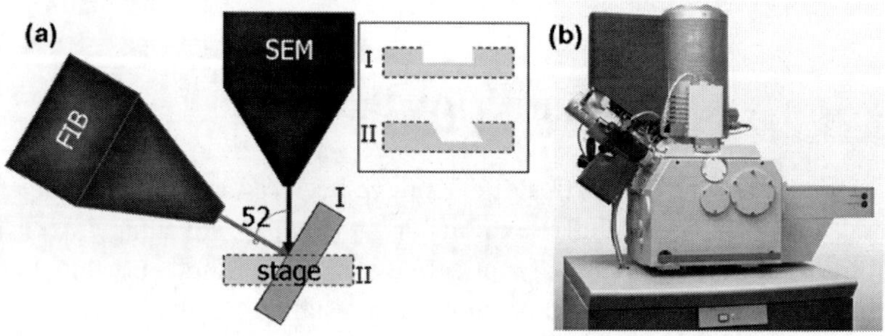

Figure 4.1. Dual Beam (SEM/FIB) machine. (a) Schematics of SEM and FIB colon positioning vertical (I) and angular milling (II) [76]. (b) Strata 400 of FEI.

All fabrication processes, excluding essential debris etching and contaminated surface cleaning, are performed *in situ* by Focused Ion Beam (FIB) patterning (30 keV with Ga ions), in a Dual Beam (SEM/FIB) machine (see in Figure 4-1a,b). Coordinate milling and monitoring by ion- and electron-microscopes, respectively, combined with precise angular stage control enables milling at an arbitrary angle (θ) to the sample normal ($0° < \theta < 52°$), as is shown in Figure 5-1a. Furthermore, FIB allows platinum and silicon-oxide deposition and sample manipulation with an omni-probe. Although FIB milling strategies have been developed for the last decades, in part for transmission electron microscopy (TEM) sample preparation [87, 88], special attention should be paid to the following issues when patterning diamond:

- The diamond etch rate is generally not tabulated for the FIB. It amounts to ~1/2 of the silicon rate for large features (<1μm), and for small ones: the etch rate depends on hole diameter, and depth. For deep holes it may be even zero (due to redeposition).
- Milling produces large amount of debris that redeposit close to the etching point.
- The focusing and astigmatism correction cannot be done on in the working area to avoid contamination.
- Low ion-beam current (28pa-93pA) required for precise patterning (~5-10nm shift in holes position) make focus and astigmatism adjustment especially difficult.
- The imaging of the working area in ion beam must be reduced to minimum: every frame reduces the membrane thickness.
- When patterning a thin membrane (150-300nm), it is locally bent by mechanical stress. The bending pattern changes during the milling due to partial stress release and static charging.
- The milling cannot be realized without charge removing (by metal deposition). For instance a ~80nm of chromium deposited layer was used at Ref. [76].

The PC holes are patterned onto a thin membrane (~200-330nm) by FIB. As an example of the present process (see Figure 4-2a), a high hole verticality, which is important for cavity confinement [75], was achieved. In Figure 4-2b,c the three-missing hole PC cavity with nearly perfect focus and stigmator adjustment of the ion-beam is demonstrated.

The main problem with a FIB-milled surface is Ga contamination. The Ga is implanted and etched simultaneously by the FIB, producing a high concentration in the vicinity of the carved surface. As a result, the luminescence of the FIB-processed diamond is found to be strongly quenched, and a charge conversion of NV^- to NV^0 is observed [75, 76]. This implanted layer has been extensively studied by Time of Flight Secondary Ion Mass Spectrometry (TOF-SIMS) in Ref. [89]. In Figure 4-3, we show the measurement of Ga concentration as a function of depth in a diamond FIB milled well (the well is 95nm deep). This implantation exhibits a very high density of Ga atoms (over 10^{22} cm^{-3}) at the ~5nm vicinity to the FIB affected

Figure 4.2. FIB milled PC patterning in high resolution SEM. (a) PC holes with high verticality and hole diameter of ~105nm [75]. (b,c) Highly homogeneous 3-missing hole cavity patterned on the membrane with ~188nm in diameter [76].

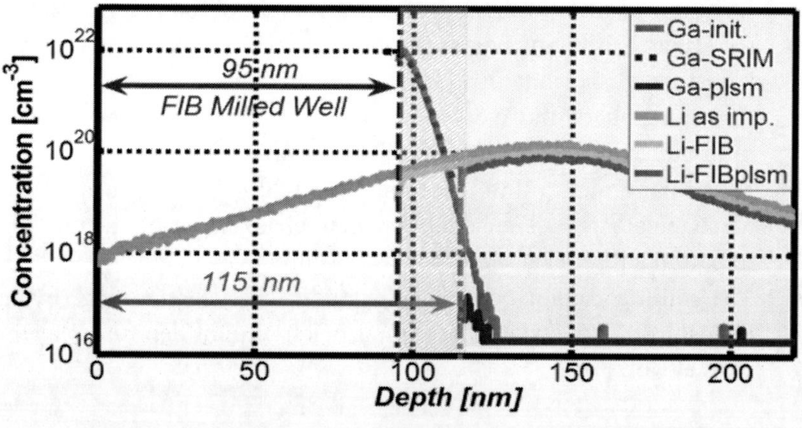

Figure 4.3. TOF-SIMS Ga and Li depth profile inside and outside 95nm deep wells FIB milled in diamond, before and after hydrogen plasma and chemical etching [89]. The diamond was implanted by 40keV Li^+ as a depth marker prior to the milling. i) as implanted Li depth profile before FIB milling (green curve) ii) Li and Ga depth profiles obtained in the FIB milled well post chemical etching (cyan and red curves, respectively) iii) Li and Ga depth profiles obtained in the FIB milled well, after H plasma and chemical etching (pink and blue curves, respectively) iiii) SRIM simulation of the FIB implanted Ga, fitting exactly the measured profile (Black dotted curve). The figure shows ~20nm damaged diamond surface etching after treatments. Note the Li profiles coincide accurately, but are slightly scaled down for an easier picture reading.

surface and then sharply decreases with increase in depth. From the photoluminescence signal (under green laser excitation), the presence of Ga can be easily discerned either by modified NV^0 emission that broadens and red-shifted followed "quenching" of NV^- centers [89] or by the presence of a graphite-like Raman signal at $\lambda \approx 580$ *nm* [90].

To remove the Ga implantation, the sample was exposed to hydrogen plasma (H_2 flow: 210 sccm, pressure: 60 Torr, MW power: 850 W, temperature: 650^0C) for 80 min with a subsequent boiling $HClO_4$: HNO_3:H_2SO_4 (1:1:1) acid mixture treatment. This treatment removes 20 (±3) nm in the Ga-milled area and improves surface roughness (RMS) by 33%. In addition, photoluminescence studies show that once the Ga rich layer is removed by H plasma, and the thus formed H termination is replaced with O termination by exposure to acids, a drastical improvement in the optical properties is noticed. Furthermore, this treatment enhances the desirable NV^- emission by a factor of ~50 with respect to unprocessed diamond. These results present a promising avenue for the realization of FIB-based diamond photonic devices with improved optical properties [89].

Another way to remove the implanted Ga is via high temperature annealing (1,000°C for 2 h in vacuum (10^{-3} Pa)) followed by a series of subsequent acid treatment (H_2SO_4:H_2O_2, 1:1, at room temperature) and oxidation in air (420°C, 2 h). This treatment has modifies diamond surface (probably by slight etching), thus facilitating the resonance tuning of the resonance peaks in optical devices [90].

Dry Etching (RIE/ICP)

The FIB based fabrication processing presented above is a very convenient device prototyping technique, and, to date, the only one that allows device fabrication *in situ*. However, serial milling process combined with high precision requirements in patterning position and uniformity needed for photonic devices requires low milling currents. This limits not only the milling speed, but also the size of photonic structures, since for sufficiently large areas the ion beam is susceptible to drifts in position, focus and stigmatism during the milling, distorting the pattern. Summing up, FIB introduces significant challenges in realization of large-scale integrated photonics. A different approach towards diamond patterning with lithographic writing accompanied by the dry etching is described in this section.

E-Beam Lithography with Reactive Ion Etching (RIE)

Electron Beam lithography (e-beam) writing presents an attractive alternative to FIB processing due to greater speed a potentially higher patterning precision. The advantage of the e-beam is based on a separation between the serial writing on an e-beam resist and parallel dry etching. This separation allows much higher throughput and may enable future interconnect of the operating cavities by a second layer aligned waveguide “masks”. The implementation of e-beam based patterning and milling in diamond has to address the following problems:

1. Resist Spinning: Typically diamond samples are relatively small compared to their Si counterparts. This makes conventional resist spinning techniques challenging due to the inhomogeneity of the resist mask at the edges.
2. Hard Mask: The e-beam resist such as Polymethyl methacrylate (PMMA) or ZEP cannot be used as a mask for diamond dry etching (they are etched much faster than diamond [91]). Thus an intermediate hard mask is required, unless a spin-on glass resist such as hydrogen silsesquioxane (HSQ) [92] or a metal layer [85] are used as the hard mask.
3. Charging: Diamond is an exceptional insulator; thus, precise e-beam writing requires a charge-removing layer, similar to the one used in FIB.
4. Development of the dry etching processing: The patterning can be transferred precisely from the e-beam resist onto the diamond only via dry etching. Usually, two different etching processes have to be developed, one for the hard mask and another for diamond, unless a negative resist layer is used.

To address these requirements, one can use ~100nm silicon oxide (SiO_2) or silicon nitride (Si_3N_4) layer, grown by chemical-vapor-deposition on a diamond surface as a hard mask for dry etching. The sample e-beam resist (PMMA or ZEP) is covered by a thin layer of metal, for instance by ~20nm thin aluminum layer (see Figure 4-4). The e-beam writing is performed via the metal, which is subsequently removed in wet-etching.

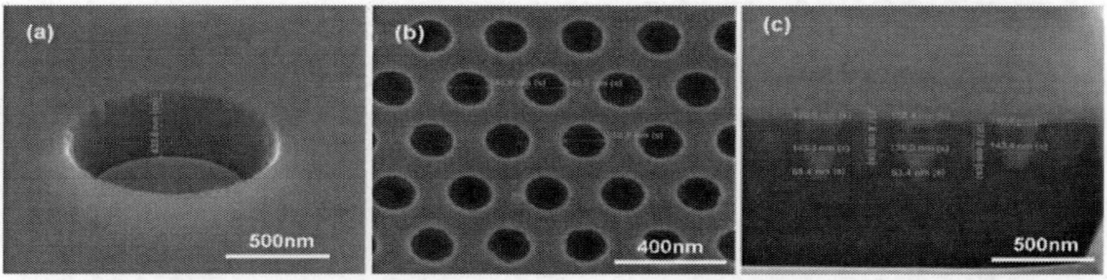

Figure 4.4. RIE patterning and etching rate SEM pictures after 26min of oxygen plasma. (a) Large aperture milling (433nm thick). (b) PC structure with average hole diameter of ~147nm. (c) Cross-section of PC row, where the hole verticality is preserved for ~220nm from the upper surface.

Then, the resist is developed and the mask (SiO_2 or Si_3N_4) is etched (for example with CHF_3 RIE), with e-beam resist acting as a mask. Afterwards, the PMMA is removed and the diamond is etched via the mask by oxygen RIE.

As one can observe from Figure 4-4, after 26 minutes of oxygen plasma, different etch rates are obtained for large and small apertures. For the holes of diameter higher than 1µm, the milling rate is ~16.5nm/min, while for small PC holes it is two times lower ~8.5nm/min. The pattern size varies slightly with the e-beam dose. Thus, etching parameters need to be recalibrated for a specific cavity design. In addition, the etching rates of the membrane are different from the bulk diamond. Therefore, additional process tuning for PC fabrication upon a diamond membrane is required.

Photolithography with Inductive Coupled Plasma (ICP)

This very similar processing can be applied to oxygen ICP. The hard mask in this case was a PECVD grown 100nm silicon-nitride (Si_3N_4) layer. This has an improved adhesion to the diamond surface, in comparison to SiO_2. The ICP has been examined for a high etching rate needed in forming waveguide or beam-like structures. The standard photolithography mask of 10µm wide waveguides was used. The silicon-nitride hard mask was removed by CHF_3 RIE, while the diamond was subjected to oxygen ICP. The results are shown in Figure 4-5.

As one can observe, the etching rate is 10 times higher (~210nm/min), as compared to the RIE; however, the sidewall quality is significantly degraded. The ICP is very efficient in removing fast "material volume", however, further RIE or FIB side-wall polishing is needed to improve the side-walls. Note, that e-beam lithography process similar to the one implemented for RIE could be realized with ICP as well.

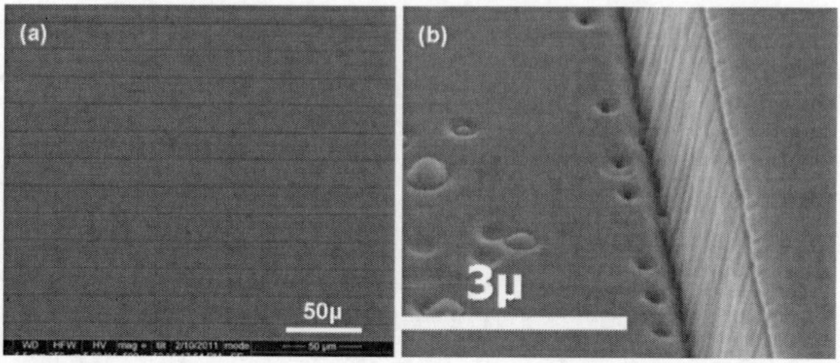

Figure 4.5. ICP milling of waveguide-like structures on diamond. (a) Low magnification SEM. (b) Close-up on the wall quality.

Additional Patterning Techniques

There are several other alternatives for the patterning process that potentially allow reducing the number of involved masks. Among them we would like to mention here two:

1. The use of a spin on glass resist such as Hydrogen silsesquioxane (HSQ) [92]. This eliminates the need of the hard mask since,after development, the resist forms a glass layer that serves as the mask for oxygen-dry etching. After etching, the mask can be removed with hydrofluoric acid. This method still requires a discharge layer on top of the resist for precise e-beam patterning.
2. A few hundred nanometer tick metallic mask such as titanium can be used in patterning diamond with ICP [85]. In this case, no discharge level is required, but the standard e-beam lithography of the metal mask is still needed.

To sum up this section, the field of diamond nano-patternig is currently extremely diversified and we believe that it will crystalize in the near future in some leading techniques that will be generally adopted by the community.

Membrane Formation

In this section, we present a brief review of several approaches towards membrane formation. These range from ion implantation, thinning of the polished/implanted membranes to nanobeam vertical and angular milling and etching.

Ion Implantation (Shallow/Deep)

The formation of a diamond slab (a membrane) separated from the substrate is a pre-requirement for the fabrication of the PC cavities and waveguides. Membrane fabrication by 'ion slicing' is based on the well-known phenomenon of the transformation of heavily ion implantation damaged diamond (in excess of some critical damage level) to graphite during annealing at high temperature [93]. In this procedure, the membrane is formed by first creating a sub-surface sacrificial layer that is subsequently etched. Ion implantation is used to form highly damaged layers at different depths and thicknesses depending on implantation conditions. We distinguish between deep and shallow implantation strategy.

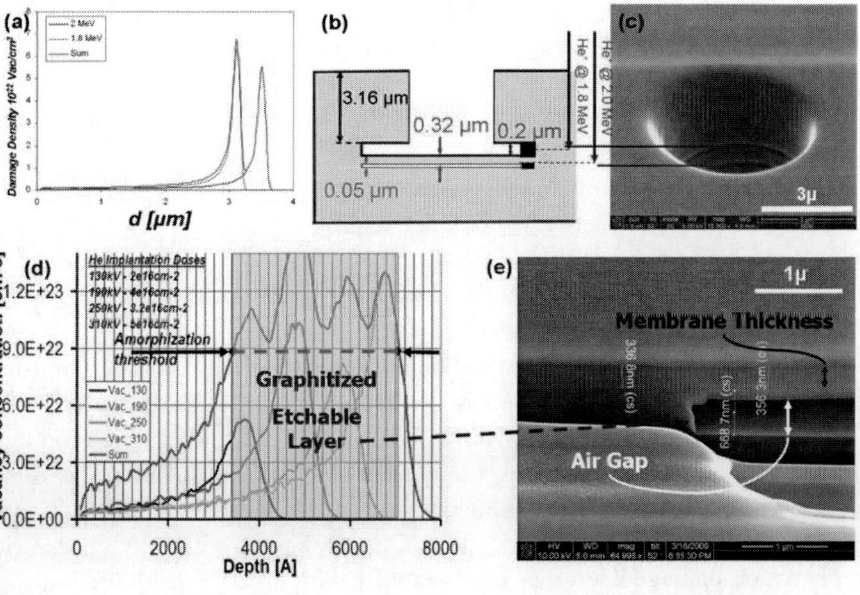

Figure 4.6. Diamond Membrane Formation [75,76]. (a,b,c) – Deep implantation processing. (a) Monte-Carle simulation of double-implantation damages [95]. (b) Schematic diagram of implanted layers. (c) Two membrane SEM view. (d,e) – Shallow implantation processing. (d) Monte-Carle simulation of four different cases of implantation damage [95]. (e) The cross-section SEM of the membrane at the stop-etch boundary.

In the deep implantation strategy, two He^+ implantations at energies of 1.8MeV and 2MeV are applied to form a thin membrane between two graphitized layers [47,48], see Figure 4-6a,b. These graphitized layers are wet-etched via FIB milled trenches that enable the penetration of the etchant. As a result, two freestanding membranes with thicknesses of 3.16μm and 0.32μm are obtained (Figure 4-6b,c). Removal of the upper 3.16μm membrane allows access and patterning of the PC structure on the thin film. One can observe from Figure 4-6b,c that beneath the lower membrane there is very small air-gap (~50nm), which can be even lower due to the membrane bending. This induces significant mode leakage into the substrate [75].

To prevent this in Ref. [94], the PC is relocated onto a pre-prepared well milled in silicon. For this purpose, a special pure-dielectric relocation technique has been developed [94].

In the shallow implantation, diamond is implanted with a multiple high dose sub-MeV He^+ ions followed by thermal annealing (see Figure 4-6d,e). As a result, a ~450nm thick buried sacrificial layer of amorphous graphite beneath the ~340nm thick mono-crystalline membrane is produced [76]. The graphitized layer is chemically etched, thus leaving a sufficient air-gap above the substrate (below the membrane) to provide optical confinement in the membrane (Figure 4-6e).

This shallow implantation technique has a significant advantage over the deep (MeV) one [75], since it does not require technologically complicated relocation procedures.

Advanced Approach to Membrane Formation

The thin membranes (200nm-300nm) formed by both shallow and deep implantations exhibit high optical losses, poor luminescence and NV^- neutralization in NV^0. [75, 76] This effect is attributed to the buried ion layer and may be rectified either by nanobeam [58] approach discussed in the following section or by introducing the advanced membrane fabrication scheme suggested in Ref. [75, 96]. This is based on the fact that ion implantation damage decreases exponentially with distance from the ion stop-range. Therefore, if the lower membrane side is subjected to dry etching removal of implantation damage via RIE or ICP, high optical grade material is expected. This fabrication strategy has been adopted at Ref. [97, 98].

Recently, progress in polishing of diamond has resulted in large single crystal diamond membranes . These are usually tens of microns thick (the numbers vary from 5 to 20). These membranes form an extraordinary material for further thinning towards ~150-250nm since they were not affected by ion-implantation. This process has been successfully demonstrated at Ref. [91, 99]. Another attractive approach towards single-crystal diamond membrane formation is based on the growth of diamond on a buffer layer of iridium/yttria-stabilized zirconia (Ir/YSZ) on silicon substrate with a subsequent removal of the initial low quality film (~11.8μm) [90].

Nanobeam Carving

An alternative way to form a vertical confinement on a bulk substrate is inclined etching (or milling). In this technique the material is etched producing angular side-walls, effectively undercutting in two opposite directions to form a constriction at the bottom of a narrow geometrical shape. This constriction can become the cusp of a freestanding beam with a triangular cross section. In the case of a diamond substrate, this process avoids the high-energy ion implantation for membrane formation. In Figure 4-7 an unpatterned beam structure produced by FIB is presented [89].

This triangular nanobeam cavity design and fabrication were first reported at [76]. There the structure was made by FIB and further studies were done to improve the optical properties of the beams by removing the damages induced by FIB [89, 96].

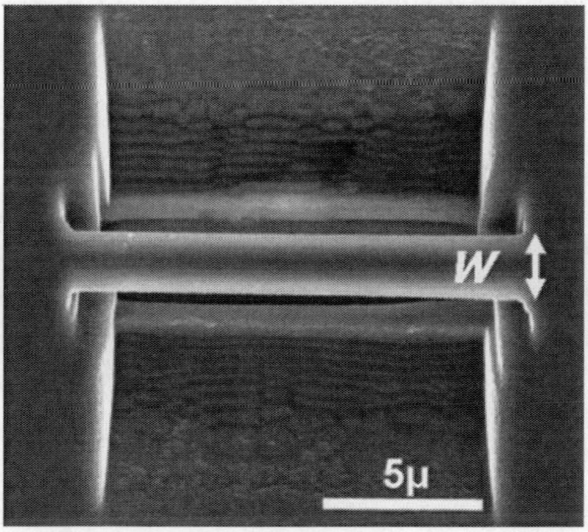

Figure 4.7. Inclined milled triangular nanobeam [94].

Recently similar structures were produced using specially developed angular ICP [85]. In this case the sample is positioned inside a triangular Faraday cage and ballistic movement of the ions penetrating into the cage provides the required inclined etching [85]. This technique combined with the triangular nanobeam cavity designs [84] may produce high-*Q* cavities from a bulk diamond material.

Surface Treatments

The following is a typical surface treatment applied in diamond processing:

- Annealing is performed at a temperature of 700°C in forming gas for 30min. This procedure induces graphitization of the ion implantation damage, once exceeding a certain critical density of broken diamond bonds,
- Chemical etching of amorphous (sp2 bonded) carbon formed by ion implantation or milling is performed either by electrochemistry in boric acid solution or by chemical etching in a boiling $HClO_4/HNO_3/H_2SO_4$ (1:1:1) acid-mix. In the later the surface is oxidized.
- Hydrogen plasma with subsequent boiling acid-mix treatment is used to remove FIB induced Ga implantation [94]. As a result, the surface is etched by ~20nm.

Formation of the NV^- by Ion-Implantation

Creating photostable and NV^- centers in the high-field region of the cavity with long spin coherence time remains an important challenge. The demonstrations of diamond cavities up to date have relied on a probabilistic approach towards this problem, i.e. fabrication of many cavities in a scarcely implanted diamond with post-selection of those that contain NV^- coupled to the cavity mode. A requirement from the NV center for QIT is that it should be *negatively charged* i.e. a NV^-; ideally, it should be an *isolated center* in the diamond, located at *pre-determined cavity center*. These pose very stringent requirement on the ways by which the NVs are introduced and are formed that are briefly reviewed below.

The accepted model for the formation of the NV^- center is the trapping of a vacancy by a substitutional nitrogen (N) atom during post-implantation annealing [100]. The vacancies are usually introduced into a diamond already containing the N impurity either from previous nitrogen ion implantation or during growth. The vacancies are formed by damaging ion implantation or by other defect formation mechanisms (i.e. by electron irradiation). Thermal treatment is expected to induce diffusion so as to form the desired NV defect. This process needs to be well under control and its yield should be as high as possible. Furthermore, the NV should have a long spin coherence time, which requires it to reside in an environment with low electromagnetic fluctuations (minimal fluctuationg charges or paramagnetic impurities).

The nitrogen impurities in the diamond can either be in the "as received" natural (type Ib) or CVD grown diamonds or introduced by N ion implantation. Sometimes implantation of the ^{15}N isotope into an ultra-pure diamond, which contains very few unintentional ^{14}N is performed. This is done to experimentally adress the implanted ^{15}NV center, and avoid the disturbance from unavoidable ^{14}NV's [101].

To control the number and location of nitrogen atoms in the sample, single N ion implantations from well-focused N ion-beams, implantations through masks or through other constrictions have been performed, with some success, as reviewed recently in Ref. [102]. Once the N ion has been introduced into the diamond matrix, it must capture a vacancy to form the NV center, preferably without losing its pre-determined location. This required the presence of vacancies in the vicinity of the N atom. Note that the vacancies formed by the implantation of a single N ion can seldom serve as a source of vacancies for NV formation. Instead, many vacancies originating from other implantations, electron or neutron irradiations must be formed in the vicinity of the N to serve as a source for capture by the N impurity. These must be driven to combine, at high efficiency, with the substitutional nitrogen by thermal annealing, yet without losing the original predetermined location of the N. This poses severe experimental problems that are still under intensive study. Different approaches to achieve efficient formation of NV centers to meet the above requirements have been published, including well-defined single N ion implantations and well-defined damage implantations. Different implantation/annealing schemes have also been applied to increase the NV formation efficiency [103]. Very recently, work describing the local controlled introduction of vacancies into well-known locations in N containing diamond by irradiation with well-focused electrons from a STM has been reported [104]. Following moderate thermal annealing (800C) of such implantations an increase of NV formation probability by a factor 1.8 has been observed. Prolonging the coherence time has also been achieved recently by controlling the implantation and annealing temperatures, thus removing perturbing defects from the vicinity of the NV center [105].

Diamond Photonics - State of the Art Review

As one reviews the progress of modern electronics and photonics, it becomes evident that they owe their advancement to extraordinary developments in material science. This has inspired new ideas in the design and architecture of novel devices. One of the main problems in diamond hinges on the fact that conventional nanofabrication techniques cannot be directly translated to diamond. Furthermore, only recently with the maturing of the ideas of quantum computing discussed above, the issue of diamond photonics has become timely. Here, we review the progress that has been made in the field during the last decade.

The first PC cavity potentially ready for NV^- coupling was realized in nanodiamond film by C. F. Wang *et al* in 2007 (see Ref. [73]). The film was grown on a silicon-dioxide/silicon substrate with PC patterning produced by electron-beam lithography via silicon-dioxide hard mask and reactive-ion-etching (RIE). The cavity showed $Q=585$ limited by the material optical losses, with no evidence of NV^-. This fabrication has illuminated the main challenge of diamond photonics for years to come: *Although, it is possible to fabricate high quality nanodiamond crystals with NV^- embedded [38], such a structure when extended to a several hundred nanometer thick polycrystalline film presents significant boundary scattering and absorption losses, with no traits on NVs.* Since then, the field has evolved in two major directions: hybrid and single crystal diamond.

In the *hybrid approach*, an NV^- is coupled to a semiconductor cavity. As a host material, silica and gallium phosphide (GaP) were chosen for their transparency in the visible spectrum. Ring resonators [106] and photonic crystal cavities [107, 108] were demonstrated. The gallium phosphide has a significant advantage due to its high refractive index ($n=3.16$) and weak parasitic luminescence [107, 108]. This approach has demonstrated a deterministic coupling between a nanocrystal-registered single NV and the PC cavity [107, 108]. Its realization benefits from relative ease in fabrication and nanocrystal manipulation, while suffering from inferior nanocrystal quantum properties [109] and inherently lower coupling between the nanocrystal and the cavity [106-108].

The single crystal diamond approach has the advantage that NVs could exhibit longer coherence time, being embedded in a larger crystal [20], while the cavity Q and NV^--cavity coupling improve significantly, potentially allowing realization of ultra-high-Q nanocavities via photonic crystal [52] or nanobeam [83, 84] geometries. Most of these implementations required the introduction of new membrane fabrication technologies, since for many years it was considered that single crystal diamond can be grown on a single crystal diamond substrate only (until buffer growth was demonstrated [90]). The usage of ion implantation for membrane formation, based on the graphitization threshold density discovered earlier [93], has been proposed to overcome this problem [110-112]. Although initial waveguiding capabilities were demonstrated [110, 111], the membrane optical losses were not quantified until the first realization of PCs nanocavities [74, 75]. This first demonstration [75] introduced Focused-Ion-Beam (FIB) based approaches for PC fabrication and analyzed the fabrication impact on the PC cavity's Q and V_m. Most importantly, it showed that for a thin membrane (~150-250nm), the implantation leads to significant optical losses and can neutralize the NV^- center. Meanwhile, several new photonic devices not relying on membrane, such as nanowire single photon sources [92] and solid immersion lenses [113] were introduced. These presented a novel inductive-coupled-plasma (ICP) etching techniques for the nanowires [92] and the first demonstration of fabrication alignment between the FIB pattern and NV^- for the lenses [113]. Furthermore, it was shown that dry-etching techniques to carve diamonds still maintain long NV^- coherence [114].

Next approaches developed in a single crystal diamond photonics were rectangular- [83] / triangular- [76] nanobeams and ring cavities [91] fabricated by FIB and e-beam lithography combined with RIE. The triangular nanobeam demonstrated the Q of ~ 220 (Ref. [76]), while the ring showed $Q\sim 4300$ (Ref. [91]). However, ring structures have high mode volume [91] ($V_m=17 \div 32\times(\lambda/n)^3$), while for the advanced cavity design of triangular nanobeam the $Q\approx 2.51\times 10^6$ and $V_m=1.06\times(\lambda/n)^3$ were predicted [84]. These are further improved by the preferential upward radiation inherent to the nanobeam design, which significantly improves the collection efficiency. Furthermore, the removal FIB-induced damages [94] produced a cavity with $Q\sim 700$ and $V_m\sim 3.6\times(\lambda/n)^3$, enabling gentle tuning of the cavity resonances [96]. This approach was further developed by introducing of angular reactive-ion-etching [85]. This allowed parallel fabrication of many beams with potentially much better precision and repeatability than by FIB, resulting in $Q=10^3$ [115].

The membrane-based technology has evolved in three different directions:

(1) The growth of a single crystal diamond membrane upon the buffer layer with a subsequent removal of the initial low quality film followed by patterning of PCs and rectangular nanobeams with FIB [90]. This showed the maximal Q of 700 for the nanobeam and the Q of ~450 for PC cavities. The cavities were coupled to silicon-vacancy centers in diamond and showed the enhancement of spontaneous emission of the factor of ~3. The reduction of the cavity Q was attributed to FIB limitation such as hole angularity [90].
(2) The implantation produced membrane with a subsequent damage layer removal by dry etching [98]. The PC cavity was produced by e-beam lithography and dry etching, while the membrane was transferred onto the PMMA layer. The PC cavity showed $Q \sim 10^3$ limited by the PMMA layer beneath the membrane.
(3) Thinning of polished membranes by dry etching [99]. The membrane was patterned by e-beam lithography and dry etching to form the PC cavities, while the underneath silicon was removed to form a free standing structure. This cavity showed the Q of 3,000 and coupled to the produced the modification of spontaneous emission by a factor of ~70. This is for now the record in producing the high-Q cavity in the field and in a maximal spontaneous emission modification of the NV.

OVERVIEW AND PROSPECTS

In this chapter we tried to give a snapshot of the state of the art diamond photonics presenting different approaches towards diamond device design and fabrication. Looking back on the last decade, we can see great advances in the technology and strong growth in the science community's interest towards the field. Only five years ago the first steps towards realization of integrated diamond photonics were taken. These focused on developing photonic crystal design and nanofabrication technologies. Now, we can see a wide variety of single devices such as lenses, wires, nanobeams and photonic crystals implemented by different fabrication approaches. This is expected to "crystallize" further into developed fabrication processing that will be stimulated by high quality diamond samples that become available and by the maturity of diamond nanotechnology.

Drawing a parallel to the word of microelectronics, one might argue that we are standing now on the verge of the era when single NV^--based quantum-bits registered into cavities are becoming available in a manner conceptually similar to the appearance of the single transistors back at 50s. Thus, the next big-breakthrough in the field will be an effort to integrate them together in attempt to form quantum-computing hardware which analogues to the first chips. The great impact in this direction may appear from the fields of sensing, since diamond based sensors may become commercially available long before the appearance of the first truly integrated quantum computer. The influence of working commercially available devices will certainly provide further impetus to develop diamond technology.

As for the realization of large scale integrated quantum hardware, although the all-in-diamond approach seems very promising, we envision that these may ultimately culminate in a hybrid chip where all-diamond cavities provide optimal coupling and extremely long spin-coherence, while being interconnected on a scalable semiconductor platform, such as gallium-phosphide; thus, benefiting from the low-loss networking between diamond qubits.

This may lead to new directions in the basic study of quantum optics and in practical devices in the fields of sensing, biology and quantum information. Summing up, diamond photonics future looks much more brilliant today than it was few years ago…

REFERENCES

[1] Michael A. Nielsen, and I. L. Chuang, *Quantum computation and quantum information* (Cambridge Univ. Press, Canbridge, 2000).

[2] D. Bouwmeester, A. K. Ekert, and A. Zeilinger, *The Physics of Quantum Information: Quantum Cryptography, Quantum Teleportation, Quantum Computation* (Springer Publishing Company, Incorporated, 2000).

[3] J. I. Cirac, P. Zoller, H. J. Kimble, and H. Mabuchi, "Quantum State Transfer and Entanglement Distribution among Distant Nodes in a Quantum Network," *Physical Review Letters* 78, 3221 (1997).

[4] A. D. Boozer, A. Boca, R. Miller, T. E. Northup, and H. J. Kimble, "Reversible State Transfer between Light and a Single Trapped Atom," *Physical Review Letters* 98, 193601 (2007).

[5] H. J. Kimble, "The quantum internet," Nature 453, 1023-1030 (2008).

[6] D. P. DiVincenzo, "*The Physical Implementation of Quantum Computation,*" (arXiv:quant-ph/0002077v3, 2008).

[7] J. L. O'Brien, G. J. Pryde, A. G. White, T. C. Ralph, and D. Branning, "Demonstration of an all-optical quantum controlled-NOT gate," *Nature* 426, 264-267 (2003).

[8] D. Leibfried, B. DeMarco, V. Meyer, D. Lucas, M. Barrett, J. Britton, W. M. Itano, B. Jelenkovic, C. Langer, T. Rosenband, and D. J. Wineland, "Experimental demonstration of a robust, high-fidelity geometric two ion-qubit phase gate," *Nature* 422, 412-415 (2003).

[9] F. Schmidt-Kaler, H. Haffner, M. Riebe, S. Gulde, G. P. T. Lancaster, T. Deuschle, C. Becher, C. F. Roos, J. Eschner, and R. Blatt, "Realization of the Cirac-Zoller controlled-NOT quantum gate," *Nature* 422, 408-411 (2003).

[10] X. Li, Y. Wu, D. Steel, D. Gammon, T. H. Stievater, D. S. Katzer, D. Park, C. Piermarocchi, and L. J. Sham, "An All-Optical Quantum Gate in a Semiconductor Quantum Dot," *Science* 301, 809-811 (2003).

[11] T. Yamamoto, Y. A. Pashkin, O. Astafiev, Y. Nakamura, and J. S. Tsai, "Demonstration of conditional gate operation using superconducting charge qubits," *Nature* 425, 941-944 (2003).

[12] A. M. Zaitsev, *Optical Properties of Diamond: A Data Handbook* (Springer-Verlag Berlin Heidelberg, 2001).

[13] F. Jelezko, and J. Wrachtrup, "Single defect centres in diamond: A review," *physica status solidi* (a) 203, 3207-3225 (2006).

[14] J. Harrison, M. J. Sellars, and N. B. Manson, "Optical spin polarisation of the N-V centre in diamond," *Journal of Luminescence* 107, 245-248 (2004).

[15] A. Gruber, A. Dräbenstedt, C. Tietz, L. Fleury, J. Wrachtrup, and C. v. Borczyskowski, "Scanning Confocal Optical Microscopy and Magnetic Resonance on Single Defect Centers," *Science* 276, 2012-2014 (1997).

[16] A. P. Nizovtsev, S. Y. Kilin, F. Jelezko, I. Popa, A. Gruber, and J. Wrachtrup, "NV centers in diamond: spin-selective photokinetics, optical ground-state spin alignment and hole burning," *Physica B: Condensed Matter* 340-342, 106-110 (2003).

[17] A. Nizovtsev, S. Kilin, F. Jelezko, I. Popa, A. Gruber, C. Tietz, and J. Wrachtrup, "Spin-selective low temperature spectroscopy on single molecules with a triplet-triplet optical transition: Application to the NV defect center in diamond," *Optics and Spectroscopy* 94, 848-858 (2003).

[18] K.-M. C. Fu, C. Santori, P. E. Barclay, and R. G. Beausoleil, "Conversion of neutral nitrogen-vacancy centers to negatively charged

nitrogen-vacancy centers through selective oxidation," *Appl. Phys. Lett.* 96, 121907 (2010).

[19] M. V. Hauf, B. Grotz, B. Naydenov, M. Dankerl, S. Pezzagna, J. Meijer, F. Jelezko, J. Wrachtrup, M. Stutzmann, F. Reinhard, and J. A. Garrido, "Chemical control of the charge state of nitrogen-vacancy centers in diamond," *Physical Review* B 83, 081304 (2011).

[20] G. Balasubramanian, P. Neumann, D. Twitchen, M. L. Markham, R. Kolesov, N. Mizuochi, J. Isoya, J. Achard, J. Beck, J. Tissler, V. Jacques, P. R. Hemmer, F. Jelezko, and J. Wrachtrup, "Ultralong spin coherence time in isotopically engineered diamond," *Nat. Mater* 8, 383-387 (2009).

[21] P. Tamarat, T. Gaebel, J. R. Rabeau, M. Khan, A. D. Greentree, H. Wilson, L. C. L. Hollenberg, S. Prawer, P. Hemmer, F. Jelezko, and J. Wrachtrup, "Stark Shift Control of Single Optical Centers in Diamond," *Physical Review Letters* 97, 083002 (2006).

[22] J. Wrachtrup, and F. Jelezko, "Processing quantum information in diamond," *Journal of Physics: Condensed Matter* 18, 807 (2006).

[23] G. D. Fuchs, V. V. Dobrovitski, D. M. Toyli, F. J. Heremans, and D. D. Awschalom, "Gigahertz Dynamics of a Strongly Driven Single Quantum Spin," *Science* 326, 1520-1522 (2009).

[24] F. Jelezko, T. Gaebel, I. Popa, M. Domhan, A. Gruber, and J. Wrachtrup, "Observation of Coherent Oscillation of a Single Nuclear Spin and Realization of a Two-Qubit Conditional Quantum Gate," *Physical Review Letters* 93, 130501 (2004).

[25] T. Gaebel, M. Domhan, I. Popa, C. Wittmann, P. Neumann, F. Jelezko, J. R. Rabeau, N. Stavrias, A. D. Greentree, S. Prawer, J. Meijer, J. Twamley, P. R. Hemmer, and J. Wrachtrup, "Room-temperature coherent coupling of single spins in diamond," *Nature Physics* (2006).

[26] P. Neumann, N. Mizuochi, F. Rempp, P. Hemmer, H. Watanabe, S. Yamasaki, V. Jacques, T. Gaebel, F. Jelezko, and J. Wrachtrup, "Multipartite Entanglement Among Single Spins in Diamond," in *Science*(2008), pp. 1326-1329.

[27] M. V. G. Dutt, L. Childress, L. Jiang, E. Togan, J. Maze, F. Jelezko, A. S. Zibrov, P. R. Hemmer, and M. D. Lukin, "Quantum Register Based on Individual Electronic and Nuclear Spin Qubits in Diamond," *Science* 316, 1312-1316 (2007).

[28] P. Neumann, R. Kolesov, B. Naydenov, J. Beck, F. Rempp, M. Steiner, V. Jacques, G. Balasubramanian, M. L. Markham, D. J. Twitchen, S. Pezzagna, J. Meijer, J. Twamley, F. Jelezko, and J. Wrachtrup,

"Quantum register based on coupled electron spins in a room-temperature solid," *Nat. Phys.* 6, 249-253 (2010).

[29] E. Togan, Y. Chu, A. S. Trifonov, L. Jiang, J. Maze, L. Childress, M. V. G. Dutt, A. S. Sorensen, P. R. Hemmer, A. S. Zibrov, and M. D. Lukin, "Quantum entanglement between an optical photon and a solid-state spin qubit," *Nature* 466, 730-734 (2010).

[30] A. D. Greentree, J. Salzman, S. Prawer, and L. C. L. Hollenberg, "Quantum gate for Q switching in monolithic photonic-band-gap cavities containing two-level atoms," *Physical Review* A 73, 013818 (2006).

[31] D. D. Awschalom, R. Epstein, and R. Hanson, "The Diamond Age of Spintronics," *Scientific American* 297, 84-91 (2007).

[32] S. J. Devitt, A. D. Greentree, R. Ionicioiu, J. L. O'Brien, W. J. Munro, and L. C. L. Hollenberg, "Photonic module: An on-demand resource for photonic entanglement," *Physical Review* A 76, 052312 (2007).

[33] S. J. Devitt, A. D. Greentree, R. Ionicioiu, J. L. O'Brien, W. J. Munro, and L. C. L. Hollenberg, "Photonic module: An on-demand resource for photonic entanglement," *Physical Review* A 76, 052312 (2007).

[34] M. S. Shahriar, P. R. Hemmer, S. Lloyd, P. S. Bhatia, and A. E. Craig, "Solid-state quantum computing using spectral holes," *Physical Review A* 66, 032301 (2002).

[35] *Cavity Quantum Electrodynamics (Advances in Atomic, Molecular and Optical Physics)* (Academic Press, 1994).

[36] A. Young, and et al., "Cavity enhanced spin measurement of the ground state spin of an NV center in diamond," *New Journal of Physics* 11, 013007 (2009).

[37] G. Balasubramanian, I. Y. Chan, R. Kolesov, M. Al-Hmoud, J. Tisler, C. Shin, C. Kim, A. Wojcik, P. R. Hemmer, A. Krueger, T. Hanke, A. Leitenstorfer, R. Bratschitsch, F. Jelezko, and J. Wrachtrup, "Nanoscale imaging magnetometry with diamond spins under ambient conditions," *Nature* 455, 648-651 (2008).

[38] J. M. Taylor, P. Cappellaro, L. Childress, L. Jiang, D. Budker, P. R. Hemmer, A. Yacoby, R. Walsworth, and M. D. Lukin, "High-sensitivity diamond magnetometer with nanoscale resolution," *Nat .Phys.* 4, 810-816 (2008).

[39] R. M. Wilson, *Nanodiamonds are promising quantum probes of living cells* (AIP, 2011).

[40] M. O. Scully, and M. S. Zubairy, *Quantum Optics* (Cambridge Univ. Press, Cambridge, 1997).

[41] E. Yablonovitch, "Inhibited Spontaneous Emission in Solid-State Physics and Electronics," *Physical Review Letters* 58, 2059 (1987).

[42] D. Englund, D. Fattal, E. Waks, G. Solomon, B. Zhang, T. Nakaoka, Y. Arakawa, Y. Yamamoto, and J. Vuckovich, "Controlling the Spontaneous Emission Rate of Single Quantum Dots in a Two-Dimensional Photonic Crystal," *Physical Review Letters* 95, 013904 (2005).

[43] S. Noda, M. Fujita, and T. Asano, "Spontaneous-emission control by photonic crystals and nanocavities," *Nat. Photon* 1, 449-458 (2007).

[44] K. J. Vahala, "Optical microcavities," *Nature* 424, 839-846 (2003).

[45] A. Boca, R. Miller, K. M. Birnbaum, A. D. Boozer, J. McKeever, and H. J. Kimble, "Observation of the Vacuum Rabi Spectrum for One Trapped Atom," *Physical Review Letters* 93, 233603 (2004).

[46] J. P. Reithmaier, G. Sek, A. Loffler, C. Hofmann, S. Kuhn, S. Reitzenstein, L. V. Keldysh, V. D. Kulakovskii, T. L. Reinecke, and A. Forchel, "Strong coupling in a single quantum dot-semiconductor microcavity system," *Nature* 432, 197-200 (2004).

[47] E. Peter, P. Senellart, D. Martrou, A. Lemaître, J. Hours, J. M. Gérard, and J. Bloch, "Exciton-Photon Strong-Coupling Regime for a Single Quantum Dot Embedded in a Microcavity," *Physical Review Letters* 95, 067401 (2005).

[48] D. W. Vernooy, V. S. Ilchenko, H. Mabuchi, E. W. Streed, and H. J. Kimble, "High-Q measurements of fused-silica microspheres in the near infrared," *Opt. Lett.* 23, 247-249 (1998).

[49] D. K. Armani, T. J. Kippenberg, S. M. Spillane, and K. J. Vahala, "Ultra-high-Q toroid microcavity on a chip," *Nature* 421, 925-928 (2003).

[50] T. Yoshie, A. Scherer, J. Hendrickson, G. Khitrova, H. M. Gibbs, G. Rupper, C. Ell, O. B. Shchekin, and D. G. Deppe, "Vacuum Rabi splitting with a single quantum dot in a photonic crystal nanocavity," *Nature* 432, 200-203 (2004).

[51] J. Vučković, M. Lončar, H. Mabuchi, and A. Scherer, "Design of photonic crystal microcavities for cavity QED," *Physical Review* E 65, 016608 (2001).

[52] I. Bayn, and J. Salzman, "Ultra high-Q photonic crystal nanocavity design: The effect of a low-ε slab material," *Opt. Express* 16, 4972-4980 (2008).

[53] S. John, "Strong localization of photons in certain disordered dielectric superlattices," *Physical Review Letters* 58, 2486 (1987).

[54] J. Joannopoulos, S. Johnson, J. Winn, and R. Meade, *Photonic Crystals: Molding the Flow of Light (Second Edition)* (Princeton University Press, 2008).

[55] C. Kittel, *Introduction to Solid State Physics* (John Wiley and Sons, Inc., 2005).

[56] S. Noda, K. Tomoda, N. Yamamoto, and A. Chutinan, "Full Three-Dimensional Photonic Bandgap Crystals at Near-Infrared Wavelengths," *Science* 289, 604-606 (2000).

[57] S. Y. Lin, J. G. Fleming, D. L. Hetherington, B. K. Smith, R. Biswas, K. M. Ho, M. M. Sigalas, W. Zubrzycki, S. R. Kurtz, and J. Bur, "A three-dimensional photonic crystal operating at infrared wavelengths," *Nature* 394, 251-253 (1998).

[58] S. G. Johnson, S. Fan, P. R. Villeneuve, J. D. Joannopoulos, and L. A. Kolodziejski, "Guided modes in photonic crystal slabs," *Physical Review B* 60, 5751 (1999).

[59] S. Tomljenovic-Hanic, M. J. Steel, C. M. de Sterke, and J. Salzman, "Diamond based photonic crystal microcavities," *Opt. Express* 14, 3556-3562 (2006).

[60] I. Bayn, and J. Salzman, "High-Q photonic crystal nanocavities on diamond for quantum electrodynamics," *The European Physical Journal - Applied Physics* 37, 19-24 (2007).

[61] S. G. Johnson, P. R. Villeneuve, S. Fan, and J. D. Joannopoulos, "Linear waveguides in photonic-crystal slabs," *Physical Review B* 62, 8212 (2000).

[62] D. Englund, I. Fushman, and J. Vuckovich, "General recipe for designing photonic crystal cavities," *Opt. Express* 13, 5961-5975 (2005).

[63] B.-S. Song, S. Noda, T. Asano, and Y. Akahane, "Ultra-high-Q photonic double-heterostructure nanocavity," *Nat. Mater* 4, 207-210 (2005).

[64] E. K., M. N., S. M., A. S., T. T., and T. W., *Ultrahigh-Q photonic crystal nanocavities realized by the local width modulation of a line defect* (AIP, 2006).

[65] Y. Takahashi, H. Hagino, Y. Tanaka, B.-S. Song, T. Asano, and S. Noda, "High-Q nanocavity with a 2-ns photon lifetime," *Opt. Express* 15, 17206-17213 (2007).

[66] T. Asano, B.-S. Song, and S. Noda, "Analysis of the experimental Q factors (~ 1 million) of photonic crystal nanocavities," *Opt. Express* 14, 1996-2002 (2006).

[67] *BandSolve 4.1 - User Manual* (RSoft Design Group, Inc., 2008).

[68] K. S. Yee, "Numerical solution of initial boundary value problems involving Maxwell's equations in isotropic media," *IEEE Trans. Antennas and Propagation* AP-14 (1966).

[69] J. Jackson, *Classical Electrodynamics Third Edition* (Wiley, 1998).

[70] M. Qiu, "Micro-cavities in silicon-on-insulator photonic crystal slabs: Determining resonant frequencies and quality factors accurately," *Microwave and Optical Technology Letters* 45, 381-385 (2005).

[71] J. P. Berenger, "A perfectly matched layer for the absorption of electromagnetic waves," *J. Comput. Phys.* 114, 185-200 (1994).

[72] C. Kreuzer, J. Riedrich-Möller, E. Neu, and C. Becher, "Design of Photonic Crystal Microcavities in Diamond Films," *Opt. Express* 16, 1632-1644 (2008).

[73] C. F. Wang, R. Hanson, D. D. Awschalom, E. L. Hu, T. Feygelson, J. Yang, and J. E. Butler, "Fabrication and characterization of two-dimensional photonic crystal microcavities in nanocrystalline diamond," *Appl. Phys. Lett.* 91, 201112 (2007).

[74] I. Bayn, B. Meyler, A. Lahav, J. Salzman, B. A. Fairchild, S. Prawer, and F. L. Martinez, "Photonic crystals (PC) in diamond: Ultra-high-Q nanocavity design, analysis and fabrication," in *ECIO*(Eindhoven, Holland 2008), p. FrC5.

[75] I. Bayn, B. Meyler, A. Lahav, J. Salzman, R. Kalish, B. A. Fairchild, S. Prawer, M. Barth, O. Benson, T. Wolf, P. Siyushev, F. Jelezko, and J. Wrachtrup, "Processing of photonic crystal nanocavity for quantum information in diamond," *Diamond and Related Materials* 20, 937-943 (2011).

[76] I. Bayn, B. Meyler, J. Salzman, and R. Kalish, "Triangular nanobeam photonic cavities in single-crystal diamond," *New Journal of Physics* 13, 025018 (2011).

[77] M. Notomi, E. Kuramochi, and H. Taniyama, "Ultrahigh-Q Nanocavity with 1D Photonic Gap," *Opt. Express* 16, 11095-11102 (2008).

[78] P. B. Deotare, M. W. McCutcheon, I. W. Frank, M. Khan, and M. Loncar, "High quality factor photonic crystal nanobeam cavities," *Appl. Phys. Lett.* 94 (2009).

[79] Y. Gong, B. Ellis, G. Shambat, T. Sarmiento, J. S. Harris, and J. Vuckovic, "Nanobeam photonic crystal cavity quantum dot laser," *Opt. Express* 18, 8781-8789 (2010).

[80] M. W. McCutcheon, and M. Loncar, "Design of a silicon nitride photonic crystal nanocavity with a Quality factor of one million for

coupling to a diamond nanocrystal," *Opt. Express* 16, 19136-19145 (2008).

[81] J. Chan, M. Eichenfield, R. Camacho, and O. Painter, "Optical and mechanical design of a ?zipper? photonic crystaloptomechanical cavity," *Opt. Express* 17, 3802-3817 (2009).

[82] Y. Gong, and J. Vuckovich, "Photonic crystal cavities in silicon dioxide," *Appl. Phys. Lett.* 96, 031107 (2010).

[83] T. M. Babinec, J. T. Choy, K. J. M. Smith, M. Khan, and M. Loncar, "Design and focused ion beam fabrication of single crystal diamond nanobeam cavities," *Journal of Vacuum Science and Technology B Microelectronics and Nanometer Structures* 29, 010601 (2011).

[84] I. Bayn, J. Salzman, and R. Kalish, "Ultra-high-Q nanobeam cavity design in Diamond," arXiv:1012.5481v1 (2010).

[85] M. J. Burek, N. P. de Leon, B. J. Shields, B. J. M. Hausmann, Y. Chu, Q. Quan, A. S. Zibrov, H. Park, M. D. Lukin, and M. Lončar, "Free-Standing Mechanical and Photonic Nanostructures in Single-Crystal Diamond," *Nano Letters* (2012).

[86] G. Balasubramanian, P. Neumann, D. Twitchen, M. Markham, R. Kolesov, N. Mizuochi, J. Isoya, J. Achard, J. Beck, J. Tissler, V. Jacques, P. R. Hemmer, F. Jelezko, and J. Wrachtrup, "Ultralong spin coherence time in isotopically engineered diamond," *Nat. Mater* 8, 383-387 (2009).

[87] L. A. Giannuzzi, and F. A. Stevie, "A review of focused ion beam milling techniques for TEM specimen preparation," *Micron* 30, 197-204 (1999).

[88] L. Giannuzzi, and F. A. Stevie, "Introduction to focused ion beams instrumentation, theory, techniques, and practice," (Springer, 2005), http://www.amazon.ca/exec/obidos/redirect?tag=citeulike09-20andamp;path=ASIN/0387231161.

[89] I. Bayn, A. Bolker, C. Cytermann, B. Meyler, V. Richter, J. Salzman, and R. Kalish, "Diamond processing by focused ion beam-surface damage and recovery," *Appl. Phys. Lett.* 99, 183109 (2011).

[90] J. Riedrich-Moller, L. Kipfstuhl, C. Hepp, E. Neu, C. Pauly, F. Mucklich, A. Baur, M. Wandt, S. Wolff, M. Fischer, S. Gsell, M. Schreck, and C. Becher, "One- and two-dimensional photonic crystal microcavities in single crystal diamond," *Nat. Nano* 7, 69-74 (2012).

[91] A. Faraon, P. E. Barclay, C. Santori, K.-M. C. Fu, and R. G. Beausoleil, "Resonant enhancement of the zero-phonon emission from a colour centre in a diamond cavity," *Nat. Photon* 5, 301-305 (2011).

[92] T. M. Babinec, B. J. M. Hausmann, M. Khan, Y. Zhang, J. R. Maze, P. R. Hemmer, and M. Loncar, "A diamond nanowire single-photon source," *Nat. Nano* 5, 195-199 (2010).

[93] C. Uzan-Saguy, C. Cytermann, R. Brener, V. Richter, M. Shaanan, and R. Kalish, "Damage threshold for ion-beam induced graphitization of diamond," *Appl. Phys. Lett.* 67, 1194-1196 (1995).

[94] I. Bayn, A. Bolker, C. Cytermann, B. Meyler, V. Richter, J. Salzman, and R. Kalish, "*Diamond Processing by Focused Ion Beam– Surface Damage and Recovery*," submitted to APL (2011).

[95] J. F. Ziegler, M. D. Ziegler, and J. P. Biersack, "*SRIM - The stopping and range of ions in matter (2010),"* Nuclear Instruments and Methods in Physics Research Section B: Beam Interactions with Materials and Atoms 268, 1818-1823 (2010).

[96] I. Bayn, "Diamond Photonic Crystals: Possible Implementation of Room Temperature Atom-Photon Interaction Experiments," in *Electrical Engineering*(Technion, Haifa, 2011).

[97] O. Gaathon, J. S. Hodges, E. H. Chen, L. Li, S. Bakhru, H. Bakhru, D. Englund, and R. M. Osgood Jr., " Planar fabrication of arrays of ion-exfoliated single-crystal-diamond membranes with nitrogen-vacancy color centers," *Optical Materials 35, 361,* (2013).

[98] J. C. Lee, A. P. Magyar, D. O. Bracher, I. Aharonovich, and E. L. Hu, "Fabrication of thin diamond membranes for photonic applications," arXiv:1210.0125 (2012).

[99] A. Faraon, C. Santori, Z. Huang, V. M. Acosta, and R. G. Beausoleil, "Coupling of Nitrogen-Vacancy Centers to Photonic Crystal Cavities in Monocrystalline Diamond," *Physical Review Letters* 109, 033604 (2012).

[100] A. Mainwood, "Nitrogen and nitrogen-vacancy complexes and their formation in diamond," *Physical Review B* 49, 7934-7940 (1994).

[101] J. R. Rabeau, P. Reichart, G. Tamanyan, D. N. Jamieson, S. Prawer, F. Jelezko, T. Gaebel, I. Popa, M. Domhan, and J. Wrachtrup, "Implantation of labelled single nitrogen vacancy centers in diamond using [sup 15]N," *Applied Physics Letters* 88, 023113-023113 (2006).

[102] S. Pezzagna, B. Naydenov, F. Jelezko, J. Wrachtrup, and J. Meijer, "Creation efficiency of nitrogen-vacancy centres in diamond," *New Journal of Physics* 12, 065017 (2010).

[103] B. Naydenov, V. Richter, J. Beck, M. Steiner, P. Neumann, G. Balasubramanian, J. Achard, F. Jelezko, J. Wrachtrup, and R. Kalish,

"Enhanced generation of single optically active spins in diamond by ion implantation," *Appl. Phys. Lett.* 96, 163108-163103 (2010).

[104] P. M. J Schwartz, C D Weis and T Schenkel, "In situ optimization of co-implantation and substrate temperature conditions for nitrogen-vacancy center formation in single-crystal diamonds " *New Journal of Physics* 13, 035022 (2011).

[105] B. Naydenov, F. Reinhard, A. Lammle, V. Richter, R. Kalish, U. F. S. D'Haenens-Johansson, M. Newton, F. Jelezko, and J. Wrachtrup, "Increasing the coherence time of single electron spins in diamond by high temperature annealing," *Applied Physics Letters* 97, 242511-242513 (2010).

[106] C. Santori, and et al., "Nanophotonics for quantum optics using nitrogen-vacancy centers in diamond," *Nanotechnology* 21, 274008 (2010).

[107] J. Wolters, A. W. Schell, G. Kewes, N. Nusse, M. Schoengen, H. Doscher, T. Hannappel, B. Lochel, M. Barth, and O. Benson, "Enhancement of the zero phonon line emission from a single nitrogen vacancy center in a nanodiamond via coupling to a photonic crystal cavity," *Appl. Phys. Lett.* 97, 141108 (2010).

[108] D. Englund, B. Shields, K. Rivoire, F. Hatami, J. Vučković, H. Park, and M. D. Lukin, "Deterministic Coupling of a Single Nitrogen Vacancy Center to a Photonic Crystal Cavity," *Nano Letters* 10, 3922-3926 (2010).

[109] J. R. Rabeau, A. Stacey, A. Rabeau, S. Prawer, F. Jelezko, I. Mirza, and J. Wrachtrup, "Single Nitrogen Vacancy Centers in Chemical Vapor Deposited Diamond Nanocrystals," *Nano Letters* 7, 3433-3437 (2007).

[110] P. Olivero, S. Rubanov, P. Reichart, B. C. Gibson, S. T. Huntington, J. Rabeau, A. D. Greentree, J. Salzman, D. Moore, D. N. Jamieson, and S. Prawer, "Ion-Beam-Assisted Lift-Off Technique for Three-Dimensional Micromachining of Freestanding Single-Crystal Diamond," *Advanced Materials* 17, 2427-2430 (2005).

[111] P. Olivero, S. Rubanov, P. Reichart, B. C. Gibson, S. T. Huntington, J. R. Rabeau, A. D. Greentree, J. Salzman, D. Moore, D. N. Jamieson, and S. Prawer, "Characterization of three-dimensional microstructures in single-crystal diamond," *Diamond and Related Materials* 15, 1614-1621 (2006).

[112] B. A. Fairchild, P. Olivero, S. Rubanov, A. D. Greentree, F. Waldermann, R. A. Taylor, I. Walmsley, J. M. Smith, S. Huntington, B. C. Gibson, D. N. Jamieson, and S. Prawer, "Fabrication of Ultrathin

Single-Crystal Diamond Membranes," *Advanced Materials* 20, 4793-4798 (2008).

[113] L. Marseglia, J. P. Hadden, A. C. Stanley-Clarke, J. P. Harrison, B. Patton, Y. L. D. Ho, B. Naydenov, F. Jelezko, J. Meijer, P. R. Dolan, J. M. Smith, J. G. Rarity, and J. L. O'Brien, "Nanofabricated solid immersion lenses registered to single emitters in diamond," *Appl. Phys. Lett*. 98, 133107-133103 (2011).

[114] J. S. Hodges, L. Li, M. Lu, E. H. Chen, M. E. Trusheim, S. Allegri, X. Yao, O. Gaathon, H. Bakhru, and D. Englund, "Long-lived NV − spin coherence in high-purity diamond membranes," *New Journal of Physics* 14, 093004 (2012).

[115] M. J. Burek, B. J. Shields, N. de Leon, B. M. Hausmann, Y. Chu, Q. Quan, M. Lukin, and M. Lončar, "Angle-etched free-standing photonic crystal nanobeam cavities in single-crystal diamond," in *CLEO*(San Jose, 2012).

In: Photonic Crystals
Editor: Donald Claus
ISBN: 978-1-62618-193-9

Chapter 2

PHOTONIC CRYSTAL FIBER NONLINEAR CHARACTERIZATION AND APPLICATION TO GENERATE MULTI-WAVELENGTH BRILLOUIN FIBER LASER

Roghaieh Parvizi[1]* and Siamak Dawazdah Emami[2]
[1]Department of Physics, Faculty of Science,
Yasouj University, Yasouj, Iran
[2]Department of Electrical Engineering, Faculty of Engineering,
University of Malaya, Kuala Lumpur, Malaysia

ABSTRACT

This chapter describes one of the main applications of photonic crystal fibre (PCF) as a nonlinear gain medium to generate multi-wavelength fiber laser based on stimulated Brillouin scattering (SBS) and four-wave mixing (FWM) nonlinear effects. Firstly, the threshold power condition of SBS for a piece of 100 m long PCF has been theoretically studied in terms of fibre parameters and input pump power. The threshold exponential gain, G_{th}, can be anticipated by this simulation which strongly depends on the fibre length Brillouin gain content and effective area. This value varies from 14 to 18 depending on the fibre length according to the Smith's definition. In order to characterize completely,

* Email parvizi1360@gmail.com.

the frequency shift of Brillouin Stokes in the PCF is measured nearly 9.75 GHz via heterodyne measurement method. To demonstrate the PCF application in fiber laser generation process, a simple multi-wavelength Brillouin fibre laser (BFL) is illustrated using the PCF in conjunction with a fibre Bragg grating (FBG) to produce at least 7 Stokes and 4 anti-Stokes lines with a line spacing of 0.08 nm. To produce double-spacing Brillouin lines, a compact multi-wavelength Brillouin fibre laser in conjunction with a figure-of-eight configuration is proposed. At a BP power of 15.3 dBm, at least 4 simultaneous lines with 20 GHz or 0.16 nm line spacing is achieved by removing the odd-order Stokes lines. The anti-Stokes lines are also generated via FWM process in the laser cavity. The output spectrum of the proposed BFL can be tuned by 80 nm, dependent on the availability of an appropriate BP source. These results show that the use of a highly nonlinear PCF is capable to produce a compact and efficient multi-wavelength laser for many applications such as wavelength division multiplexing system (WDM) optical communication network system and sensors.

1. INTRODUCTION

As described in the last chapters, Photonic crystal fibers (PCFs) exhibit unique and dramatic nonlinear optical effects due to their periodic microstructure which forms core and cladding [1, 2]. For instance, one of the most interesting classes of PCFs is composed of a small-scale solid core with multiple air holes typically arranged in a hexagonal lattice around the core acting as the cladding. With the combination of a large refractive index contrast between the silica core and the air field microstructure, PCFs can be designed to enable tight mode confinement that results in a low effective mode area and thereby high nonlinear characteristics. Nonlinear effects in PCFs have been reported in many recent literatures which show that the fiber can achieved dramatic nonlinear effects, even with short fiber lengths [3-6]. Stimulated Brillouin scattering (SBS) is one of the important nonlinear effects in optical fibers. This effect is attributed to high intensity light in the fiber's core that vibrates molecules and then generates acoustic waves. The excited acoustic waves will introduce periodic modulation of refractive index, which can be seen as a moving grating. This grating produces a backward propagating frequency shifted Stokes and anti-Stokes components called SBS. This scattering has a significant influence on the operation of optical transmission systems as well as applications in distributed fiber sensors using narrow-linewidth single frequency lasers [7]. Therefore, many recent works have been reported on the development of an efficient Brillouin fiber laser (BFL) [8, 9]. This chapter describes one of the main applications of photonic crystal fibre (PCF) as a nonlinear gain medium to generate multi-wavelength fiber laser based on stimulated Brillouin scattering (SBS). firstly we have numerically and experimentally characterized the nonlinear features of PCF in terms of providing SBS effect by the most common and basic configuration. The power budget requirements (threshold pump power), the achievable gain and saturation characteristics of Brillouin amplifiers, precise and quantitive treatment of the SBS process is evaluated. Then, the compact configurations to provide optimized condition of Brillouin scattering and multi wavelength generation are explored which describe as single, double, odd (even) order, and multi wavelength Brillouin fiber laser respectively. The exploring of this chapter is by a photonic crystal fiber with the shown scanning electron micrograph (SEM) image of the PCF. The PCF used has a triangular core with average diameter of 2.1± 0.3 μm, cladding diameter of 128 ± 5 μm, and an effective core area A_{eff} of 6.1 μm^2. The average air hole diameter of the fiber is 0.8 μm with 1.5 μm pitch. The PCF is made from pure silica with 17.4 wt% of Ge-doped core region. The Ge-doped core functions to increase the nonlinear refractive index of the core, creates a smaller mode field diameter and reduces the confinement loss. This effect can reduce acoustic loss and avoid the broadening of Brillouin spectrum linewidth. PCF can be spliced to an intermediate fiber and then a single mode fiber (SMF) with a splice loss of 0.35 dB at each ends.

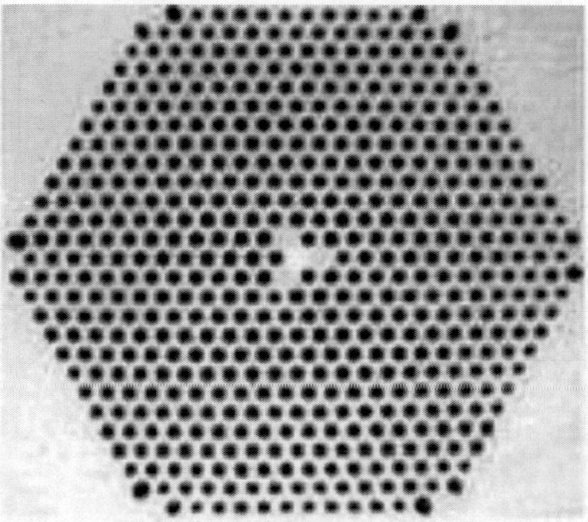

Figure 1. SEM image of the incorporated PCFR configurations.

2. Performance of Stimulated Brillouin Scattering in Optical Fibers

The acoustic phonons involved in the Brillouin scattering and the interactions occurs typically over a narrow order of tens of MHz of Brillouin gain linewidth, such as 35 MHz for fused silica, with the downshifted frequency nearly 10 GHz from the Brillouin pump wave in the wavelength 1550 nm [10-12] The shift is determined by the velocity of the acoustic grating along the fiber and is therefore dependent on the mechanical properties of the fiber such as the elasto-optic coefficient, applied strain and ambient temperature [13, 14] The Brillouin Stokes (BS) propagating in the opposite direction of a pump signal creates significant distortion within the single. In the other hand, SBS provides gain in the opposite direction of the direction of the Brillouin pump wave source. Therefore it depletes the pump power and converts it to a potentially strong Brillouin Stokes wave propagating backward to the pump direction. Consequently, the Brillouin Stokes wave must be shielded by an isolator. The simplest proper structure for generating SBS is illustrated in Figure 2 where BS wave can be monitored in the backward direction in the free end of fiber. Incorporating optical circulator in this configuration is attributed to two roles that this equipment can play. Firstly, the Brillouin pump laser source is isolated and protected by the optical circulator from any backward reflected power and secondly the reflected power is readily monitored by the optical spectrum analyzer (OSA) at the port 3 of the optical circulator.

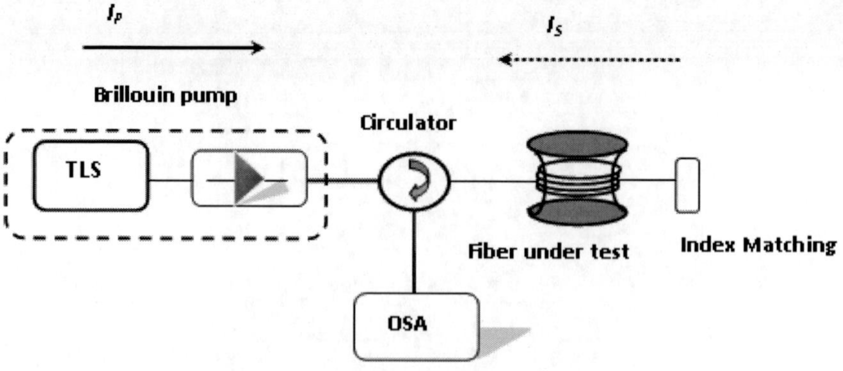

Figure 2. Experimental setup for generating Brillouin scattering in the backward direction.

In this configuration as shown in Figure 2, a tunable laser source (TLS) is used as the Brillouin pump (BP) signal generator with maximum peak power approximately 5 dBm at the wavelength 1550 nm and the BP wide linewidth about 15 MHz. Erbium doped fiber can be incorporated to amplify the TLS signal power for some cases that the higher power is needed in Brillouin fiber lasers structures. In another words, an Erbium doped fiber (EDF) amplifier (EDFA) is utilized to boost the TLS output power. As a result, the combination of the TLS and EDFA is the Brillouin pump (BP) which has the maximum output power about 15 dBm. The 14 m EDF used in the experiment has an ion concentration of 440 ppm, a numerical aperture of 0.21, and a cutoff wavelength 920 nm. The EDF is pumped by a laser diode at the pump wavelength of 980 nm with the maximum output power of 90 mW. By increasing the BP power, the evolution of the anti-Stokes, the BP reflection, and the Brillouin Stokes waves can be observed.

The maximum BP power is launched into the free end fiber through the ports 1 and 2 of the optical circulator (OC). The reflection is monitored through the port 3 of the OC by using the optical spectrum analyzer (OSA) with the resolution 0.01 nm. In Figure 3, the launched Brillouin pump and the reflected intensity spectra are illustrated as the comparison. The anti-Stokes, the BP Rayliegh reflection and the Brillouin Stokes are detected in the backward direction spectrum with the power of around -54 dBm, -5.1 dBm, and -30.5 dBm, respectively.

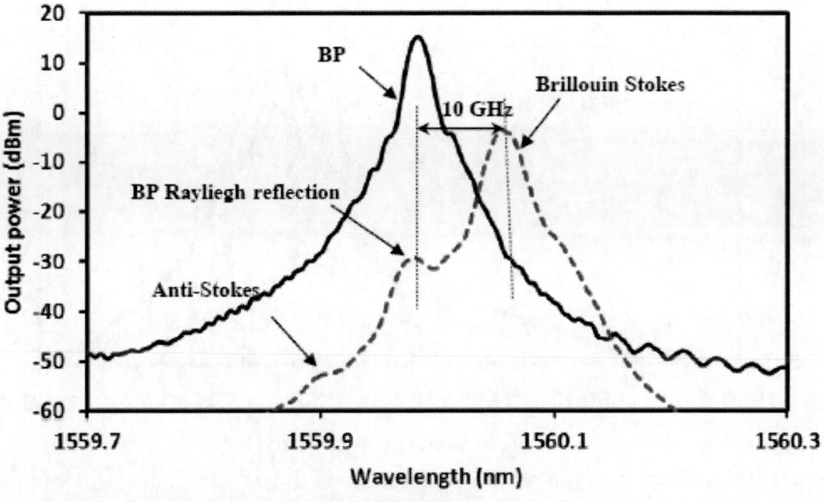

Figure 3.The launched Brillouin pump (BP) and the reflection.

The Brillouin Stokes light wavelength is shifted upward by 0.088 nm from the BP wavelength whereas the anti-Stokes wavelength is downshifted by the same spacing caused by the degenerate four-wave mixing effect happening between the Brillouin pump and Brillouin Stokes waves. The providing Brillouin gain media can be a gas, a liquid, or any solid with applications in high-power solid-state master oscillator power amplifier (MOPA) systems which provide average output powers up to kilowatt range with diffraction-limited beam quality [15].

According to the SBS theory, the beat frequency between the Brillouin scattered wave and the incident BP light is equal to the frequency of the acoustic wave. As long as the Brillouin laser pump and the Brillouin Stokes wave interactions reinforce the acoustic waves, it amplifies the scattered Brillouin Stokes wave so that the stimulated Brillouin scattering (SBS) occurs. The Brillouin threshold power is defined as the Brillouin pump power at which Brillouin Stokes power suddenly increases. One method to reduce the Brillouin threshold power is the waveguide structures utilizing which have been used in optical fibers [16-18] and in capillaries [16] where a high intensity is achieved along the full length of fiber. However, if very narrow waveguide diameters are used, damage of the incident surface and volume damage inside the capillary are the limiting factors for the maximum power [18].

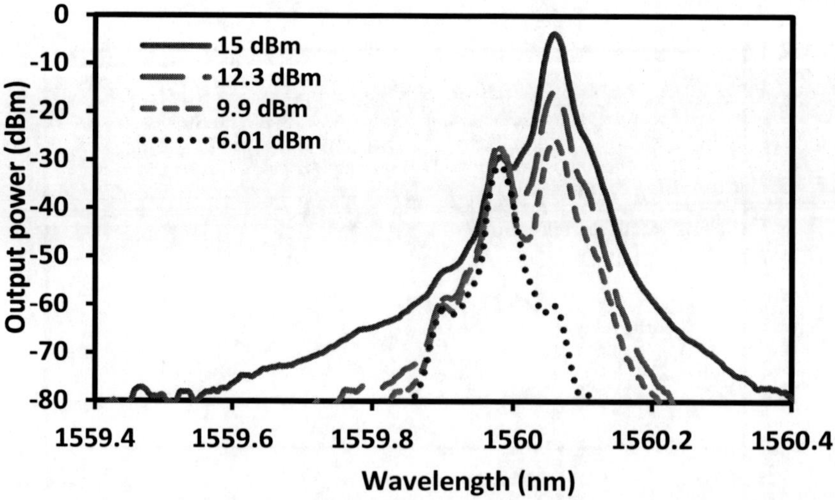

Figure 4. The evolution of the reflection spectrum is shown versus the Brillouin pump (BP) power.

The results are shown in Figure 4 in which the reflection spectra including the peak power of the anti-Stokes, the BP Rayleigh reflection, and the Brillouin Stokes are demonstrated versus BP powers with the maximum pump power of 15 dBm. By increasing BP power to 15 dBm, the Brillouin Stokes power is boosted by 51 dB due to the SBS nonlinear phenomena, whereas the anti-Stokes and the BP Rayleigh reflection peak powers are increased by only about 5 dB and 11 dB, respectively.

3. NUMERICAL EVALUATION OF BRILLOUIN THRESHOLD POWER

Brillouin scattering has been implemented in a wide range of applications as a narrowband fiber laser source [19], narrow band filtering, and Brillouin fiber amplifier (BFA). Since the Brillouin frequency shift is dependent on the material composition, the temperature and pressure of the medium, the BFA can also be applied in various optical fiber sensors. In a development of distributed Brillouin sensors (DSB), the Brillouin shift is measured along the fiber length based on frequency (or time) domain modulation of the input pump wave [19]. One of the main characteristics of the nonlinear Brillouin scattering effect in optical fibers is its threshold power condition for the occurrence of the SBS. The threshold power is normally defined as the input pump power, P_{th}, at which the output Stokes is some fraction (r), used as threshold criterion, of P_{th} [18]. This threshold input pump power value is strongly depended on the providing Brillouin gain medium or fiber length which plays an important role in the distributed Brillouin sensors accuracy. In another words, a uniform resolution measurement over the entire sensing length cannot be achievable in case of strong pump depletion [19]. Consequently, a precise and comprehensive theoretically measurement of the threshold input pump power with considering the pump depletion as well as sensing fiber length variation is required. In this section, the Brillouin scattering process in a piece of PCF is theoretically studied for possible use in a development of a Brillouin fiber laser in DSB. In this study, the Stokes emission is considered to be initiated by spontaneous Brillouin scattering, SpBS, and the 1% criterion is used for the threshold definition. The threshold power is investigated based on PCF parameters, which determines the amount of transmitted pump and Stokes powers. This investigation also involves the experimental verification.

3.1. Theoretical Principle

If the input pump intensity is sufficient, SBS in optical fibers will convert a pump field with frequency ω_p to a scattered counter-propagating probe field with frequency ω_s (Stokes waves), through an electrostrictive process. Indeed, Brillouin scattering arises from the nonlinear interaction of launched light with propagating density waves or acoustic phonons. Since momentum and energy can be conserved during these scattering events, the SBS effect is remarkable [20]. The interaction between the three fields of input pump signal, Brillouin scattered wave, and acoustic wave (phonon) along a single mode fiber of length can be represented by a set of differential equations as follows [21]:

$$\frac{dI_S}{dz} = -g_B I_P I_S + \alpha_S I_S \quad (1)$$

$$\frac{dI_P}{dz} = -\frac{\omega_P}{\omega_S} g_B I_P I_S - \alpha_P I_P \quad (2)$$

where is the Brillouin gain coefficient, I_P and I_S are the pump and Stokes wave intensities, α_p, α_s is the absorption coefficients at pump and Stokes frequency, which are denoted by ω_P and ω_S respectively and z is the position along the length of the fiber. Equation (1) is obtained using the fiber losses due to Stokes wave determined by fiber loss coefficient at Stokes frequency and counter propagating nature of this wave. The feedback process responsible for Brillouin scattering is controlled by these two coupled equations, (1) and (2).

If we assumed that $\omega_p \approx \omega_s$ and $\alpha_p \approx \alpha_s \equiv \alpha$ to describe the SBS effect based on the fact that the Brillouin shift is relatively small, equations (1) and (2) can be rewritten as [10];

$$\frac{dI_S}{dz} = -g_B I_P I_S + \alpha I_S \quad (3)$$

$$\frac{dI_P}{dz} = -g_B I_P I_S - \alpha I_P \quad (4)$$

The first terms of equation (3) and (4) represent the Brillion gain and the pump power depletion, respectively. In case the Stokes power is much smaller than the pump power, one can assume that the pump power is not depleted and therefore the first term of equation (4) can be neglected. Thus, a more familiar closed-form expression of the SBS threshold power can be obtained as follow [10]:

$$P_{th} \approx 21 \frac{A_{eff}}{g_B^{(0)} L_{eff}} \tag{5}$$

where is the effective length of interaction.

Equation (5) is one of the approaches to define SBS threshold in optical fibers. It is based on the condition in which the reflected Stokes optical power at the beginning of the fiber ($z = 0$) equals the input pump power $(P_{reflected} = P_{in})$ in the absence of pump depletion. Practically the SBS threshold power is identified as the amount of input power producing a reflected power that is equal to a fraction, r, of the pump ($P_{reflected} = r \times P_{in}$). This fraction factor or threshold criterion can vary between and according to various literatures [18]. Actually, for accurate determination of the threshold condition, the influence of pump power depletion term should be considered. Thus, considering pump depletion factor for a proper threshold definition, the simulation of the Brillouin scattering process in a fiber such as photonics crystal fiber (PCF) is revised by returning to equations (3) and (4). It should be noted that the boundary conditions, which are known as the launched pump and Stokes intensities, are specified at the opposite ends of the fiber and small enough as initial spontaneous Brillouin scattering at the fiber end. So, the algorithm adopted in solving the equations with these boundary conditions is a precise model based on the fourth order Runge-Kutta algorithm using 10^{-3} accuracy.

3.2. Numerical and Experimental Results Comparison

Figure 5(a) shows an experimental set-up for the SBS threshold measurement. The Brillouin scattering generation in a piece of fiber is schematically illustrated in figure 5(b). A PCF is used as a Brillouin gain medium in both the experimental measurement and theoretical calculation.

The backward signal composed of Rayleigh and spontaneous Brillouin scattering (P_S) is generated when the Brillouin pump (BP) is injected into the fiber. In the measurement setup, a tunable laser source (TLS) is used as a narrow linewidth laser with a maximum peak power of approximately 5 dBm at a fixed operating wavelength of 1550 nm. The pump wavelength is amplified by Erbium doped fiber amplifier (EDFA) to achieve a maximum power of around 21 dBm before it is launched into the PCF through ports 1 and 2 of the optical circulator. The reflected signal is observed and measured through port 3 of the OC using the optical spectrum analyzer (OSA) with a resolution of 0.01 nm. The PCF used as the gain medium has a triangular core with an average diameter of 2.1 μm and cladding diameter of 128 μm. A scanning electron micrograph (SEM) image of the PCF is shown in the inset of Figure 5(a).

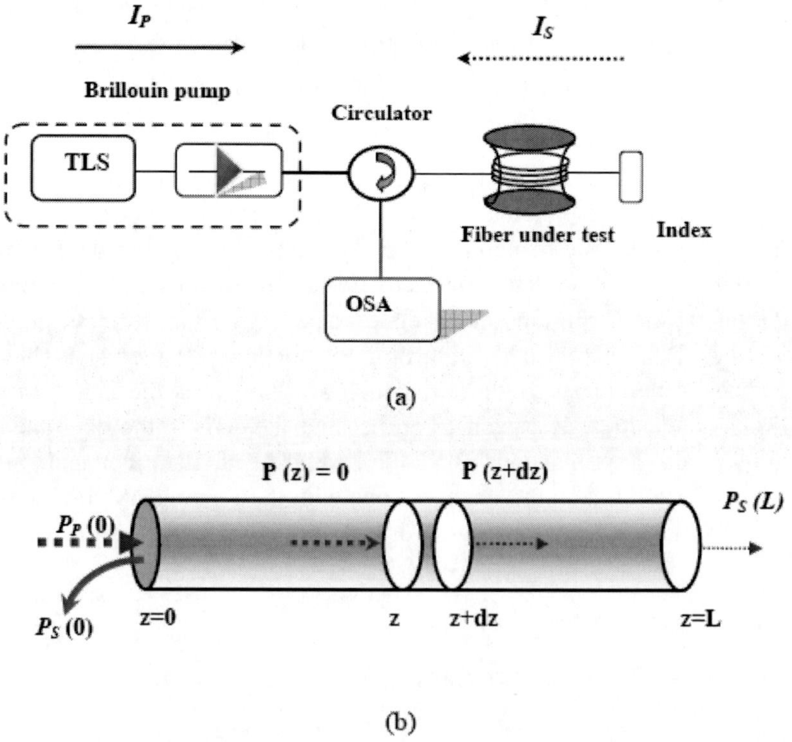

Figure 5. (a) Schematic plot of Brillouin scattering generator process. (b) The experimental configuration to measure SBS power threshold.

The average air hole diameter of the fiber is 0.8 μm with 1.5 μm pitch. The PCF length is 100 m with an effective area A of 6.1 μm^2 and an attenuation coefficient of less than 9 dB/km or α (km^{-1}) =α (dB/km) × (ln (10)/10) =2.07 km^{-1}. To explore the numerical results for a given length of the single mode fiber, L, we first assume that a pump wave with intensity of $I_p(0)$ is launched into the fiber, which initiates the Brillouin scattering in the fiber core (Singh *et al.*, 2007). The Brillouin gain coefficient, g_B, is considered to be 1.93×10^{-11} m/W at room temperature. Based on these parameters, we iteratively solve the equations over a range of values of the launched pump power $I_P(0)$ to determine the reflected Stokes intensity and its conversion efficiency. Figure 6 depicts the experimental and the numerically simulated results of the transmitted pump and backscattered Stokes power as a function of the incident Brillouin pump power. In reality, the experimental value of the backscattered power is the sum of the Rayleigh and Brillouin scatterings. Below the line of 20 dB of *Pp*(0), which represents the criterion of 1% input pump power, Rayleigh scattering is dominant whereas in the neighborhood above this line, Brillouin scattering is the main contributor to the backscattering wave output. The threshold determination by 1% criterion as illustrated in Figure 6 shows that Rayleigh scattering doesn't play a significant role in dictating the threshold value.

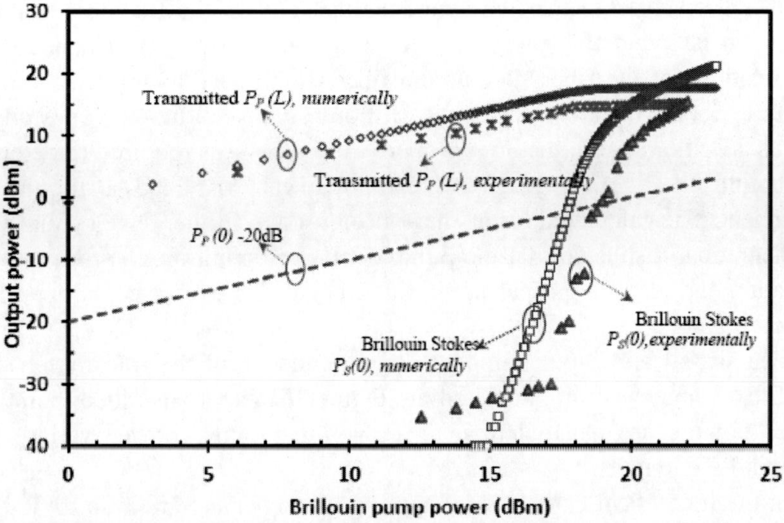

Figure 6. Plot of the transmitted pump and backscattered Stokes power obtained by the theoretical and experimental results against the input pump power.

Rayleigh scattering has been disregarded in the simulation of the Brillouin scattering power resulting in the difference between the numerical and experimental results for the Brillouin pump power of less than 15 dBm. Ideally, the SBS threshold is likely to be higher still considering other sources of loss in the experimental configuration. Nevertheless, the result shows a good agreement between the experimental outcome and the numerical prediction.

3.3. Threshold Pump Power Evaluation

Spontaneous Brillouin scattering or Stokes wave is generated as the BP is launched into a PCF due to thermal fluctuations, which modulates the refractive index. If an external laser signal interacts with the Stokes waves in a Brillouin gain medium, the stimulated Brillouin scattering (SBS) initiates and this is known as the SBS generator process. To fulfill the threshold condition for the occurrence of the SBS generation, the single-pass gain obtained by the Brillouin gain medium, $G=g_B I_P(0) L_{eff}$, must exceed the Brillouin gain which is provided by the threshold input intensity. The Brillouin gain at threshold condition is fixed at 21 in accordance with the Smith's definition which assigns 19 for single-mode-fibers. Based on the Brillouin gain interpretation which is derived from the two equations that govern the Brillouin scattering process, it is found that the threshold pump power strongly depends on the specifications of the fiber such as the fiber length used as the Brillouin gain medium, fiber's linear loss and the Brillouin gain coefficient. To study the PCF as a Brillouin gain medium properly, calculation is required to determine the optimized fiber length with certain linear loss and Brillouin gain coefficient. The calculated results have been used to further develop and verify the theoretical results so that the pump depletion condition, the Stokes power saturation and also the proper input pump power for every special length can be incorporated.

The spatial and input pump power dependence of the transmitted pump and Stokes intensity are depicted in Figure 7. The transmitted pump and Stokes powers are calculated against the input pump power variation for different PCF lengths.

In the theoretical calculation, input pump power is varied up to 300 mW using a step size of 10 mW and the PCF length is varied up to 400 m using 20 m increment. The amount of transmitted pump power at the end of the PCF is dependent on the absorbed light by the fiber and the light converted to backscattered Stokes waves. In the case where pump depletion is not considered, the transmitted power should rise linearly as the input pump power increases.

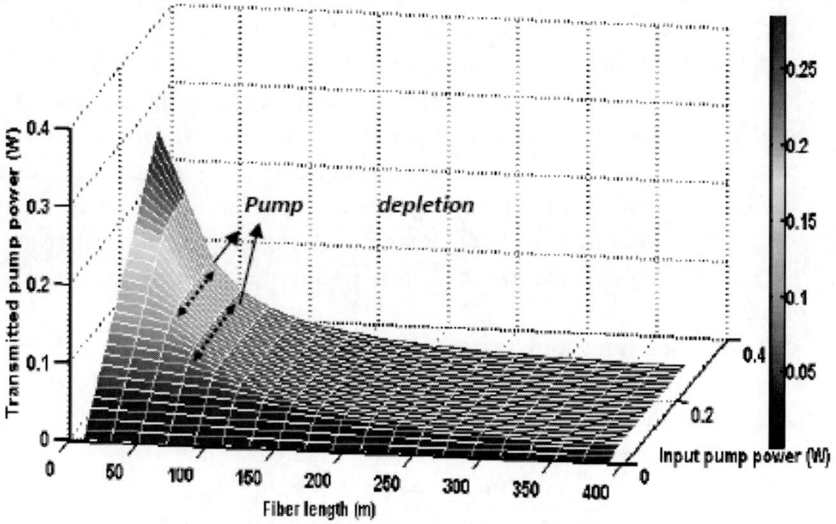

(a)

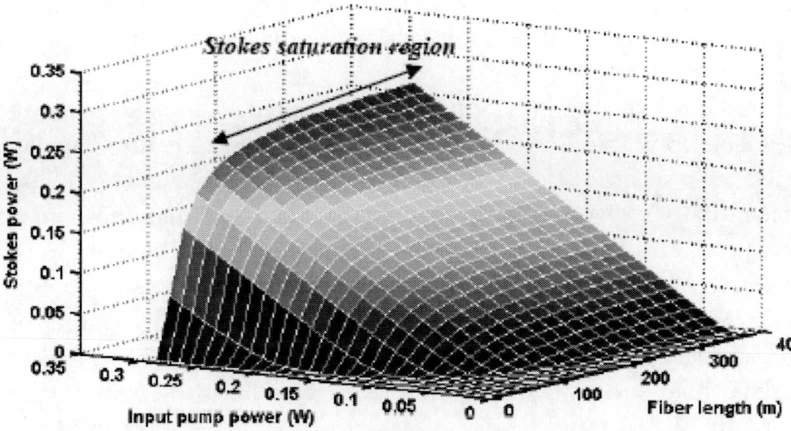

(b)

Figure 7. Plots of (a) transmitted pump and (b) backscattered Brillouin Stokes power against input pump power for different fiber length.

In DSB application, the optimized length of the sensing fiber can be determined based on the pump depletion as shown in Figure 7(a). The transmitted pump power is smaller for a longer fiber used. This is attributed to the fact that the longer length allows the pump to interact more with the gain medium to produce a higher gain. This contributes to a more efficient Brillouin Stokes conversion, which decreases the transmitted pump power. For a shorter length of fiber especially in a region of less than 50m, the SBS threshold requirements cannot be fulfilled and the pump power cannot be converted to the Stokes even though the incident pump power is set at the maximum value of 300 mW as shown in Figure 7. Also, it can be observed that the Stokes power increases with input pump power especially at a longer PCF used. The Stokes power also increases with the increment of fiber length especially at high input signal power before it saturates at a certain power level as shown in Figure 7(b).

To assess the threshold power based on various criterion, the earlier results on the transmitted pump and the backscattered Stokes powers (depicted in Figure 8) is transformed into another useful form defined as conversion efficiency percentage, $(I_S(0)/I_P(0) \times 100)$. Figure 8 shows the conversion efficiency against both input pump power and the PCF length. Equation (5) is obtained based on the condition that the reflected Stokes optical power is equal to the input pump power, which is equivalent to 100% conversion efficiency. One can also define the SBS power threshold to be the input power where the reflected power is equal to some fraction of the pump. Figure 8 shows that the conversion efficiency begins to grow sharply at around 1% against input pump power.

Then, the threshold power of the PCF is defined as an input pump power which is able to provide 1% conversion efficiency at each fiber length. The PCF used has a specific effective core area, $A_{eff} = 6.1\ \mu m^2$, and Brillouin gain coefficient of 1.93×10^{-11}m/W. This analysis shows that, the threshold pump power depends strongly on the fiber length as indicated in Figure 9. As shown in the figure, the threshold derived from measurements based on Smith's definition as well as the threshold calculated from the simulations of the coupled intensity equations (3 and 4) indicate similar trends. It is also evident that the threshold power decreases with the increase in fiber length. If the fiber length is divided to two sections of less and more than 100 m, the sharp decrease of threshold pump power trend can be seen for the length of less than 100 m. Consequently, to find out the optimized Brillouin gain against fiber length, 100 m is considered the optimum length in this special case of photonic crystal fiber.

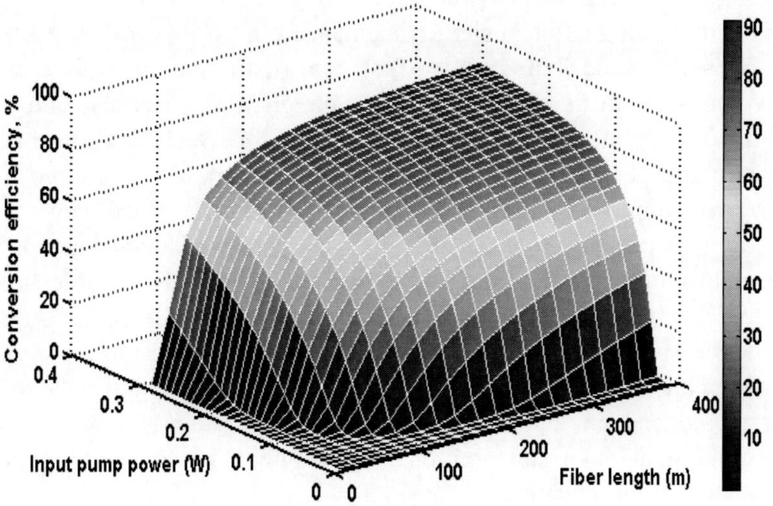

Figure 8. Plot of conversion efficiency percentage against input pump power for different length of fiber.

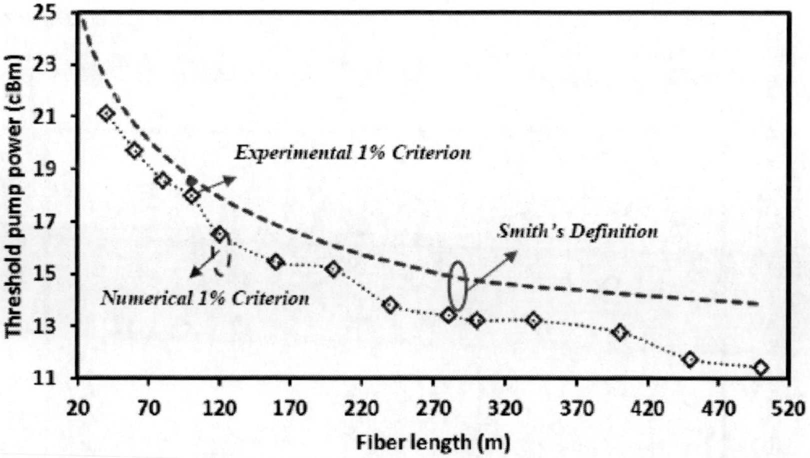

Figure 9. Plot of the threshold pump power versus fiber length, which was obtained by using 1% criterion.

It is worth noting that the Brillouin Stokes source which is placed at $z = L$ arises from spontaneous Stokes intensity considered to be 10^{-9} times of the input pump power in this simulation. Smith's estimation is basically higher than both the simulation and experimental results.

This can be attributed to two facts that firstly, the pump depletion hasn't been considered in the Smith's calculation and secondly, the threshold condition is defined as the input pump power for which this power equals the backscattered Stokes power at z = 0 based on Smith's definition. In reality, these conditions are not experimentally observable since pump depletion inevitably occurs and consequently the backscattered Stokes power is less than the input pump power. As can be observed in figure 8, the threshold power decreases with the PCF length until a convergence is reached. Corresponding to the value of A_{eff} and g_B, this fiber length is changing to provide sufficient power. This trend shows that beyond a certain length, the threshold power doesn't depend on the fiber length anymore. Figure 10 shows the plot of threshold gain as a function of PCF length. The threshold gain can be calculated from equation (5) as, $G_{th} = g_B\ P_{th}\ L_{eff}/A_{eff}$. Foremost, G_{th} is commonly taken to be a constant of nearly 21, independent of any effective SBS characterizing parameters. However in this simulation, G_{th} can be calculated precisely according to the obtained values shown in Figure10. As discussed earlier, the threshold power depends on fiber parameters and thus, G_{th} is also treated similarly. Thus, it is evident that G_{th} is not a constant as commonly assumed in the literature.

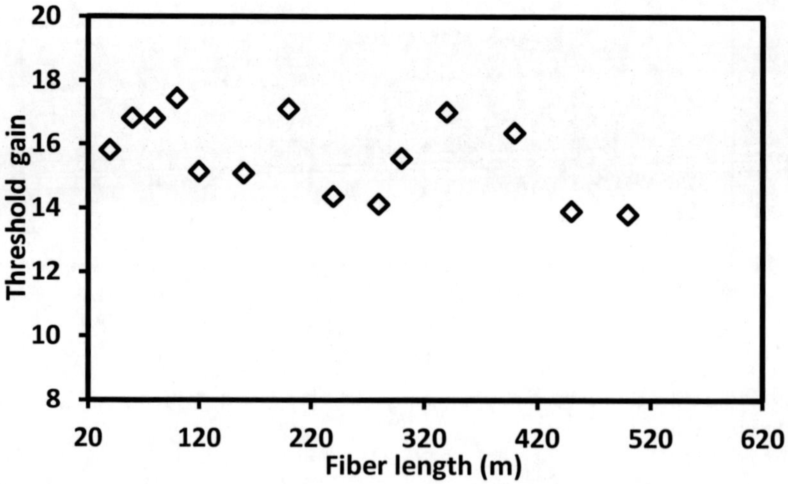

Figure 10. Plot of G_{th} against fiber length corresponding to the calculated threshold powers.

The use of higher values of g_B and smaller values of effective area provide the threshold power requirement with a shorter fiber length. Therefore, the actual value of G_{th} is fluctuating between 14 and 18 depending on the fiber length. Since P_{th} is nearly constant around 14 dBm as shown in Fig. 5 for the length of 240 to 400 m and according to this relation of $G_{th}= g_B P_{th} (L_{eff}/A_{eff})$, the fraction of (L_{eff}/A_{eff}) is considered to play the main role of evaluating G_{th}. Therefore, the higher value of G_{th} is anticipated for longer length as long as threshold pump power doesn't vary.

4. Generation of Brillouin Fiber Laser Based on Conventional Ring Cavity

One of the most prominent and suitable methods to reduce the SBS threshold is the feedback schemes in the form of ring cavities and linear cavities. Provided that Brillouin amplification overcompensates the cavity loss, Brillouin Stokes waves oscillates and it generates Brillouin fiber lasers (BFLs). The linear and ring cavity configuration makes difference in their propagation process. The linear cavity causes two counter-propagating BFL signals at the laser wavelength whereas in ring cavities, only one unidirectional BFL beam can launch in the backward direction regarding the BP direction. In the next section, the generation of BFL in a ring cavity will be discussed with different length of PCF to explore the fiber length's effect on threshold pump power. Two schemes of ring cavity configuration have been proposed to find the enhanced results. Then a compact linear cavity to generate multi wavelength BFL is studied based on double pass configuration by utilizing fiber Brag gratings instead of the mirrors. The experimental setup for the conventional ring cavity BFL is illustrated in Figure 11 which consists of an optical circulator, a 3dB coupler and a piece of PCF. The Brillouin pump (BP) is an external cavity tunable laser source (TLS) with a line-width of approximately 20MHz, which is amplified by an erbium-doped fiber amplifier (EDFA) to provide sufficient power for this study. The BP wavelength is fixed at 1560 nm in this experiment. The BP is launched through the optical circulator to generate a backward propagating SBS, which oscillates inside the resonator to generate the Brillouin laser at wavelength downshifted by 0.08 nm from the BP wavelength. The clockwise oscillating laser is extracted from the resonator via the 3dB coupler to route into the optical spectrum analyzer (OSA) via optical circulator.

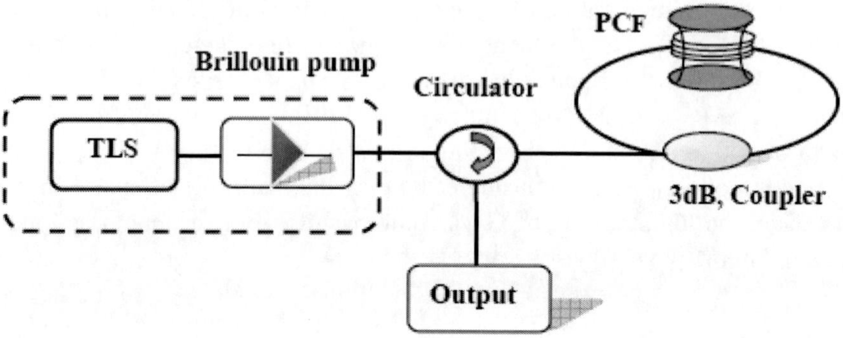

Figure 11. Experimental configurations of the PCF-based BFL conventional ring cavity.

In this BFL, only single-wavelength of the Brillouin Stokes line is obtained because the Brillouin Stokes cannot act as subsequent BP to create higher order Brillouin Stokes signals.

The output of the BFL is characterized using an OSA with a resolution of 0.015 nm. The experiments are carried out for three different lengths of PCF; 100m, 50m and 20m. Figure 12 demonstrates the output spectrum of the BFL at different lengths of PCF, 100m, 50m, and 20m while the BP power is fixed at 15.7 dBm. The line spacing is obtained at approximately 0.08 nm in the wavelength domain and 11GHz in the frequency domain, as measured by the OSA. The 3dB spectral bandwidth of the BFL is measured to be less than 0.02nm, limited by the OSA resolution.

As shown in Figure 12, side mode suppression ratios (SMSRs) of the BFL are obtained at 26.4 dB and 22.2 dB with 100 m and 50 m long PCF respectively. At 100m PCF, the peak power of the laser is obtained at more than 0dBm. Although the longer PCF in the BFL resonator increases the cavity's loss, it provides a better nonlinearity characteristic to achieve an efficient SBS using a micro-structured cladding region with air holes to guide light in a pure silica core.

The nonlinearity coefficient of the fiber is approximately $11 Wkm^{-1}$ at 1550nm region. Since the 20m PCF is not able to efficiently reach enough nonlinearity in this resonator as shown in Figure 12, its Brillouin peak doesn't provide threshold power. Figure 13 shows the measured peak power for the Stokes at different BP powers. As shown in the figure, the peak power increases with the BP power.

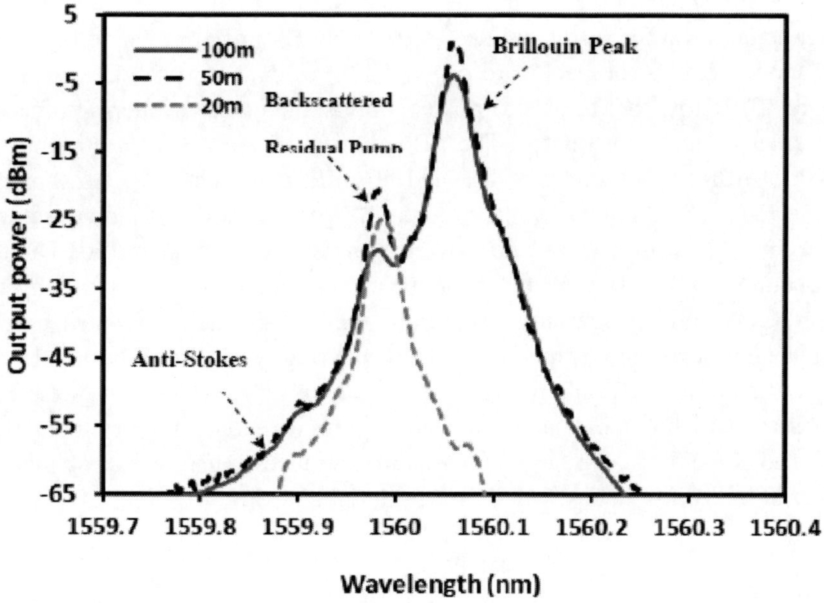

Figure 12. Output spectra of the ring BFL with different long PCF.

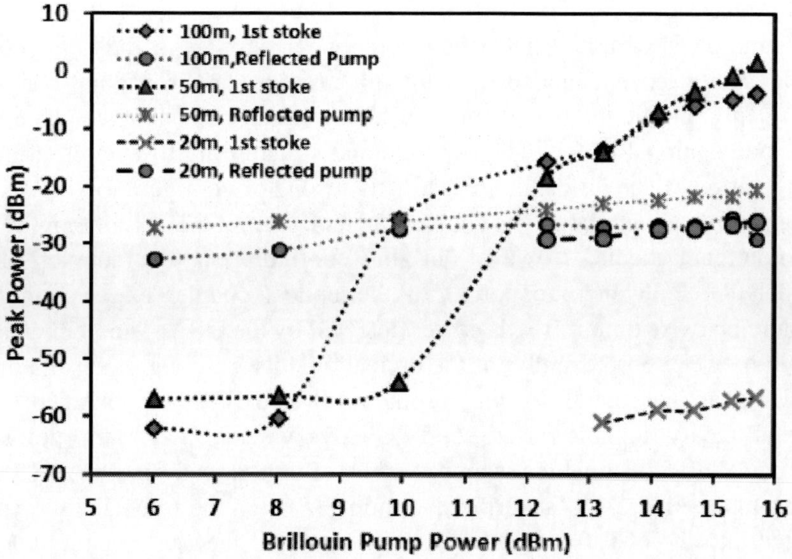

Figure 13. Brillouin Stokes power as a function of BP power at different gain medium.

The threshold power is the BP power at where the resonator starts to generate laser via SBS. It is defined as the input BP power at which the BFL's peak power equals the backward propagating residual BP power as shown in Figure 13. Below this threshold power, only spontaneous scattering emission is oscillating in the resonator. The SBS thresholds are obtained at about 9.9 and 12.3 dBm with the use of 100 and 50 m PCF respectively. As shown in Figure 13, the SBS power increases as the BP power rises and starts to saturate at around 13 dBm due to the power transfer to the higher order Stokes meanwhile for the 20 m PCF the threshold requirements are not met by this pump power and configuration. Therefore, there is also no abrupt change from spontaneous scattering to the stimulated scattering as shown in Figure 13 due to insufficient interaction length in the 20 m long PCF. The output of the BFL is observed to be stable at room temperature with only minor fluctuations observed coinciding with large temperature variances. Therefore, the threshold power of the BFL reduces with PCF length rising.

5. Generation of Brillouin Fiber Laser Based on Modified Ring Cavity

In this study, a modified Brillouin fiber ring laser (BFRL) cavity configuration is demonstrated using components similar to the conventional BFRL cavity except added isolator to provide unidirectional backward propagating, but the higher output BFRL power can be obtained as compared to the conventional BFRL. The experimental structure for the modified BFRL configuration is shown in Figure 14. This resonator configuration consists of the same optical circulator, the coupler, an isolator and the 100 m long PCF as Brillouin gain media. However, in this configuration, the coupler placed between the PCF and the port 3 of the optical coupler extracts only the counterclockwise output BFRL power detected by the OSA. The PCF which is spliced between ports 2 and 3 of the optical circulator through the isolator and coupler generates the BFL ring resonator. Both BFRLs are pumped by an external cavity tunable laser source (TLS) with a maximum power of approximately 5 dBm which is amplified by erbium doped fiber up to 15 dBm power. In the Figure 14, the Brillouin pump (BP) with the 15 MHz linewidth is shouted into the PCF from port 1 through port 2 of the optical circulator in a clockwise direction. The backward-propagating SBS generated resonates by oscillations inside the ring in counter clockwise direction to generate the

Brillouin laser, which is coupled out using a 90% coupler. The optical circulator acts as an isolator to protect the BP from any backward reflection and also force the unidirectional operation of the laser in the ring cavity made by the incorporated isolator. Although Brillouin scattering in the PCF propagates only in the backward direction based on the SBS feature.

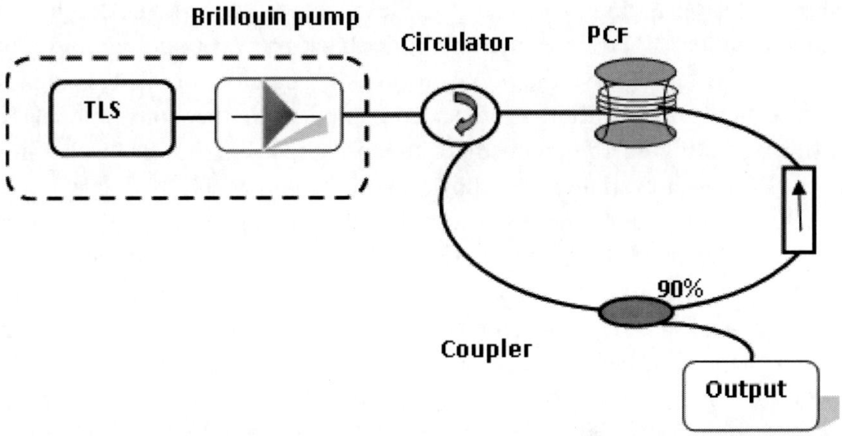

Figure 14. Experimental configurations of the PCF-based BFL modified ring cavity.

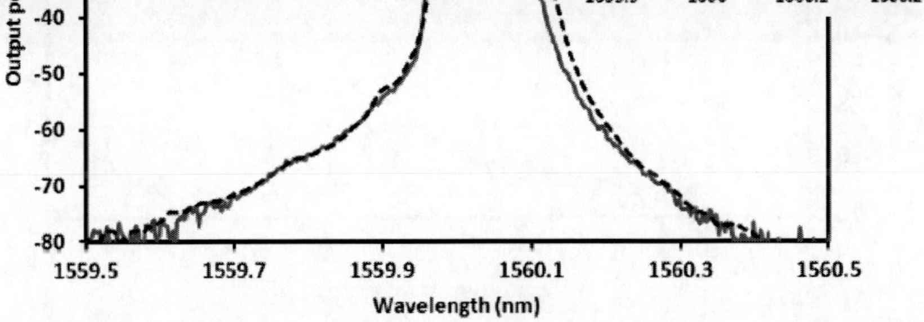

Figure 15. The Comparison of the BFRL output spectrum of the modified and conventional configurations.

The results are illustrated in Figure 15 which compares the output spectrums for the conventional BFRL and the modified BFRL configuration. Both BFRL configurations generate a Brillouin laser at about 1550.1 nm, which is 0.08 nm shifted from the BP wavelength. The modified BFRL configuration of Figure 14 shows SMSR of around 31.5 dB, which is nearly 6.3 dB higher compared to the conventional configuration in Figure 12. Since the PCF receives more BP power in the modified ring cavity rather than the conventional one, the cavity can provide a higher gain media. In the other hand, the higher BP power generates higher back-propagating Brillouin Stokes power which can improve the output spectrum. Therefore the BFL is higher in the modified BFL configuration as compared with the conventional BFL configuration. Figure 15 demonstrates the measured peak power for the Stokes against BP powers variation. As shown in this figure, generally the peak power increases with the launched BP power. The thresholds for the Brillouin Stokes are obtained around 12.7 and 10.37 dBm with the use of conventional and modified ring cavity configuration, respectively. The threshold power is the BP power at where the resonator starts to generate laser via SBS and abrupt changes happening which discussed in section 3.2. Below this threshold power, only spontaneous scattering emission is oscillating in the resonator. As shown in Figure 16, besides the higher achieved SMSR, the SBS threshold power is less for the used modified ring cavity rather than conventional one.

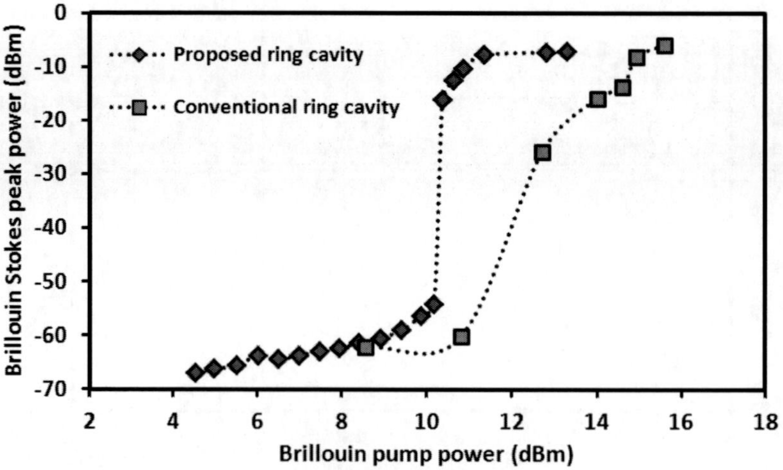

Figure 16. Brillouin Stokes power as a function of BP power by modified and conventional ring cavity configuration.

6. PCF Brillouin Frequency Shift Measurement Based on Heterodyne Method

An efficient method for linewidth, spontaneous Brillouin gain, Brillouin frequency shift measurement is the heterodyne method. Different setups are employed for different linewidth regimes. Improved laser sources recently indicate narrow linewidth in the sub-megahertz range with enhanced coherence properties. Among all fiber lasers, the linewidth of Brillouin fiber lasers has been experimentally determined in various ranges from 3.84 Hz to 2.8 kHz [22]. The ultra-narrow linewidth measurement is a challenge because the This method requires another laser either with a comparable well-known spectrum or with an extremely narrow and ignorable linewidth which can be provided by tunable laser source with the narrow linewidth of nearly 20 GHz. As discussed in chapter 2, Brillouin laser frequency is shifted with the value of nearly 10 GHz from the injected Brillouin pump wave which this value differs based on the used fiber material. In this thesis, heterodyne method has been used to compare the Brillouin frequency shift of PCF with 17.4 wt% of Ge-doped core region and silica based SMF to explore the material effect in fibers. The proposed fiber laser architecture for the measurement of Brillouin frequency shift is depicted in Figure 17, using the heterodyne method.

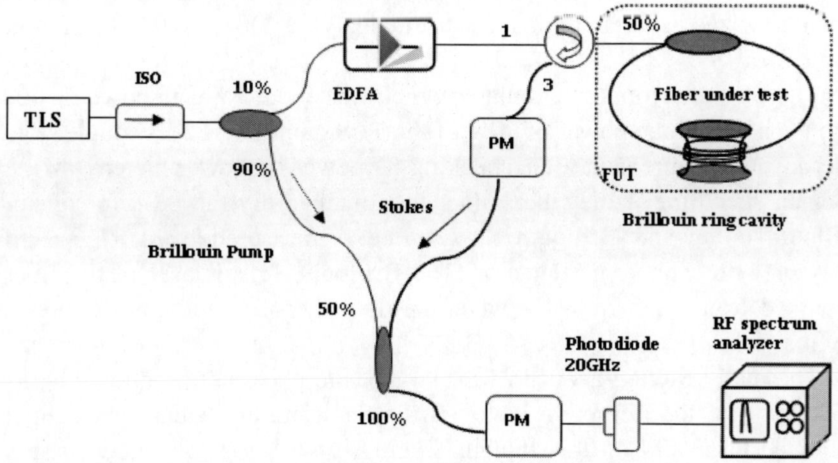

Figure 17. The experimental setup for Brillouin frequency shift measurement of a tunable laser source (TLS) by using a Brillouin fiber laser (BFL) in the heterodyne technique.

Different kinds of single mode fiber are incorporated as 25km SMF and 100m long of photonic crystal fiber (PCF). The PCF is made from pure silica with 17.4 wt% of Ge-doped core region. It is spliced to an intermediate fiber and then a single mode fiber (SMF) with a splice loss of 0.35 dB at each end. The threshold power of this PCF in backward direction is more than 19 dBm as mentioned in section 3.2; however, the power of injected signal is about 14 dBm. So to reduce the threshold power, we incorporated the ring cavity to achieve Brillouin Stokes scattering from the backward direction of the right end of the setup.

This proposed structure consists of two main sectors. A ring cavity as a Brillouin gain block which includes a piece of single mode fiber and a 3 dB coupler to provide backward propagating and Stokes line oscillation. The other sector is a heterodyne method to measure the Brillouin scattering gain. The BP is an external cavity tunable laser source (TLS) with a line-width of approximately 20MHz, which splits to two portions by a 9:1 coupler in which the 10% part is amplified by an erbium-doped fiber amplifier to provide sufficient power for this study. This amplified signal is launched through the optical circulator acting as an isolator to direct the propagation of Brillouin Stokes (BS) line in clockwise direction. Brillouin Stokes line which oscillates inside the resonator to generate the Brillouin Stokes at wavelength downshifted by 0.08 nm (nearly 10 GHz) from the BP wavelength. Before combination of the Brillouin pump and Brillouin signal, the output of the BFL is characterized using an optical spectrum analyzer (OSA) with a resolution of 0.015 nm.

The BS light from the circulator which was about −5 dBm power is mixed with the nearly same power of 90% of the TLS output by a 3dB coupler for the heterodyne measurement after checking its power by power meter (PM). The BS gain spectrum of the fiber under test obtained by the beating signals of Brillouin pump and Brillouin Stokes line is measured by a RF spectrum analyzer with a photonic detector. The Brillouin Stokes laser in the fiber is generated from interactions between the optical mode and acoustic-modes in the fiber core.

Thermally excited acoustic waves (acoustic phonons) produce a periodic modulation of the refractive index. Brillouin scattering occurs when light is diffracted backward on this moving grating, giving rise to frequency shifted Stokes and anti-Stokes components. This process can be stimulated when the interferences of the laser light and the Stokes wave reinforce the acoustic wave through electrostriction. Since the scattered light undergoes a Doppler frequency shift, the Brillouin shift depends on the acoustic velocity and is given by [23]

$$\upsilon_B = \frac{2nV_a}{\lambda} \tag{1}$$

where is the acoustic velocity within the fiber, is the refractive index and the vacuum wavelength of the incident lightwave. The strong attenuation of sound waves in silica determines the shape of the Brillouin gain spectrum (BGS). Actually, the exponential decay of the acoustic waves result in a gain presenting a Lorenzian spectral profile (Heiman, 1979).

$$g_B(\upsilon) = g_0 \frac{(\Delta\upsilon_B/2)^2}{(\upsilon-\upsilon_B)^2 + (\Delta\upsilon_B/2)^2} \tag{2}$$

where $\Delta\nu_B$ is the full-width at half maximum (FWHM). The BGS peaks at the Brillouin frequency shift ν_B (or the frequency difference between the two beams), and the peak value is given by the Brillouin gain coefficient

$$g_B(\upsilon) = g_0 = \frac{2\pi n^7 p_{12}^2}{c\lambda_p^2 \rho_0 V_a \Delta\upsilon_B} \tag{3}$$

where p_{12} is the longitudinal elasto-optic coefficient, ρ_0 is the density, λ_p is the pump wavelength and c is the vacuum velocity of light [21]. As shown in figure 17, the peak power of the back scattered Brillouin line of fiber under test after passing through the port 3 of the circulator measured by power meter. This scattering can be widely used in fiber distributed sensing, since this process is sensitive to temperature and strain linearly [24]. After controlling the power and output spectrum of the backward Brillouin laser and also Brillouin pump power, they are combined through a 3 dB coupler to provide BP and BS waves beating signal. The output frequency spectrum of the beating signals are demonstrated in Figure 18 with applying 100 m long PCF and 25 km SMF. The frequency shift of Brillouin scattering provided by PCF is nearly 9.75 GHz whereas it is around 10.83 GHz by employing SMF. This change is attributed to the 17.4 wt% of Ge-doped core region of PCF which makes shorter the Brillouin frequency shift.

7. Multi Wavelength Generation Based on Brillouin Fiber Laser With Photonic Crystal Fiber

7.1. Dual-Wavelength Fiber Laser Generation

Dual-wavelength fiber lasers have been of interest for recent years because of their potential applications in optical communications, fiber optical sensors and microwave photonics systems. In order to achieve narrow linewidth and single longitudinal mode output, various techniques have been proposed and demonstrated, including introducing an ultranarrow dual-transmission-peak bandpass filter in linear and ring cavity configurations [25, 26], cascaded distributed feedback fiber lasers, the superimposed chirp FBGs (SCFBGs) in the photosensitive erbium–ytterbium co-doped fiber [27]. Another way to achieve the dual-wavelength lasing is using a passive triple-ring cavity and a hybrid gain medium. But the parameters of the laser need to be properly adjusted to guarantee the operation [28].

A multi-wavelength lasing can also be achieved using a stimulated Brillouin scattering (SBS) process in single-mode fiber (SMF) [29]. SBS has potential applications in sensors [30] and in the realization of very narrow linewidth Brillouin fiber lasers (BFLs) to achieve dual- or multi-wavelength lasing, whereby the Stokes linewidth becomes significantly narrower than the pump linewidth with reductions of up to a few hertz [31].

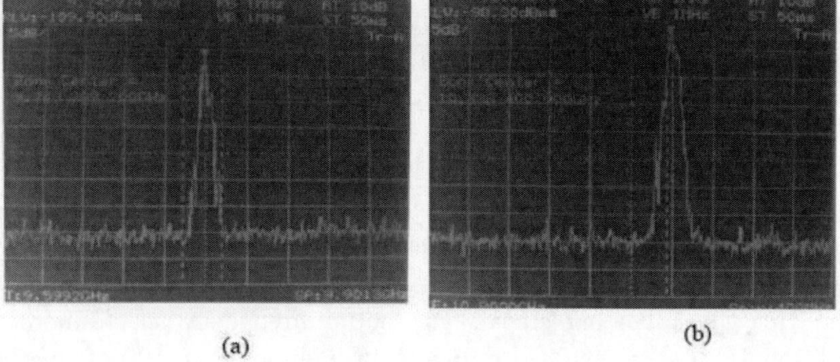

(a) (b)

Figure 18. Obtained Brillouin gain spectrum results by 100 m PCF, (a), and 25 km SMF, (b), with illustration of Brillouin frequency shift as a zone center in the graphs.

Moreover, a BFL can be designed for virtually any wavelength, provided that the required pump laser is available, thus showing a high potential for applications in metrology and spectroscopy. Recently, the emergence of photonic crystal fibers (PCF) with high nonlinearity has enabled the SBS to happen more easily in a relatively short length of PCF (usually around 100 m) with low optical power (Ahmad, et al., 2009). To date, PCF-based BFLs have been demonstrated using various configurations for both single wavelength and multi-wavelength operations [23].

In this section, a dual-wavelength configuration of a PCF-based BFL is experimentally investigated using simple dual-pass amplification architecture. The proposed laser uses a piece of PCF as a Brillouin gain medium and a broadband fiber Bragg grating (FBG) operating in C-band region as a reflector to allow dual-pass operation to achieve a wavelength tunable and stable room-temperature lasing. This type of laser does not require a linear gain medium such as EDF to increase the effect. Therefore it can operate at any wavelength depending on the Brillouin pump (BP) wavelength.

7.2. Dual Wavelength Generation Experimental Configuration

The experimental setup of the proposed dual wavelength laser is depicted in Figure 19, with the inset showing the scanning electron micrograph (SEM) image of the PCF.

It consists of a short piece of PCF, and a high-reflectivity broad band FBG at the cavity end. The FBG has a 40nm bandwidth centered at 1545nm with a reflectivity of more than 99%. An external cavity tunable laser source (TLS) with a line-width of approximately 20MHz is amplified by an erbium-doped fiber amplifier (EDFA) to provide sufficient power for this study and acts as Brillouin pump (BP). The broad band FBG operating in C band region at the end of this linear structure allows dual-pass Stokes lines oscillation in the system.

The BP is launched into the PCF through an optical circulator to generate the first order Brillouin Stokes in clockwise direction at wavelength downshifted by 0.08nm from the BP wavelength. When the laser power exceeds the threshold condition, it will then create the second order Stokes signal in the opposite direction. The residual BP and first Stokes lines are reflected by the broadband FBG to double propagating inside the PCF and oscillates in the linear cavity due to spurious reflection as well as scattering effects.

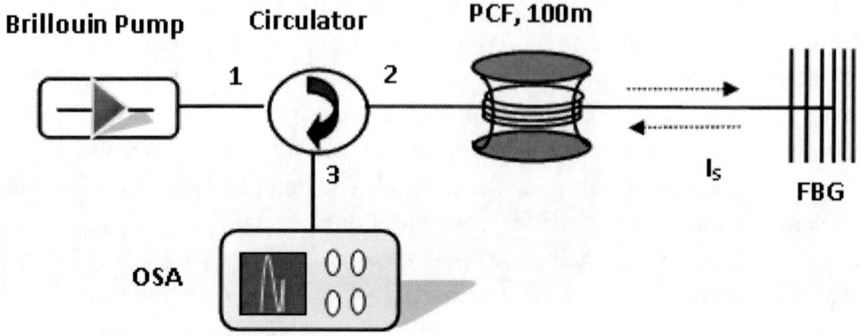

Figure 19. The proposed dual-wavelength BFL architecture.

This will increase the nonlinear gain in the resonator which in turn improves the dual-wavelength lasing process. The output of the BFL is characterized using an optical spectrum analyzer (OSA) with a resolution of 0.015 nm.

7.3. Results and Discussion of the Proposed Dual-Wavelength Configuration

Figure 3.20 shows the output spectrum of the dual wavelength laser with and without the broadband FBG. The experiment was also carried out for the configuration without the FBG to investigate the characteristic of the back-scattered light due to Brillouin and Rayleigh scatterings. In the experiment, the BP power and wavelength are fixed at 15.7 dBm and 1559.99 nm, respectively. Without the FBG, spectrum measured at the port 3 of the circulator shows a new wavelength at 1560.07 nm with a frequency shift of 0.08 nm which is a result of Rayleigh backscattering of the first-order Brillouin Stokes generated from the BP in the PCF as the Brillouin gain medium. The peak power of the Brillouin Stokes wavelength measured by the OSA in single-pass architecture is -26.9 dBm, which is 22.3 dB lower compared to that of the back-scattered BP signal. This indicates that an amplitude-equilibrium dual-wavelength lasing cannot be achieved in the single-pass amplification configuration. In dual-pass architecture, the forward Brillouin Stokes and residual BP are reflected back by the FBG and oscillates in the linear configuration (due to spurious reflection or scattering at the input part of the system) to generate dual-wavelength output.

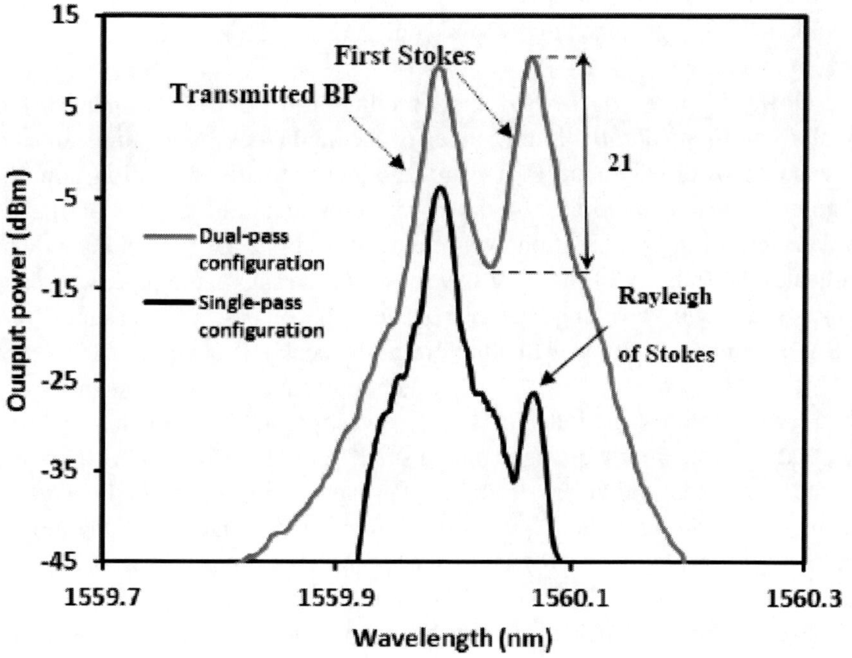

Figure 20. Output spectra of the BFL in single-pass and dual-pass configurations.

Figure 20 shows two simultaneous lasing lines at 1560.07 nm and 1560.15 nm with a wavelength separation of 0.08 nm at a BP power of 15.7 dBm when the FBG is incorporated at the end of linear system. The optical signal to noise ratio of each line is over 21 dB. By applying the FBG, the amount of oscillating BP power in the PCF is increased and therefore more power is converted to the Stokes. The Stokes power is also improved due to the increased Brillouin gain in the dual-pass system. As shown in the spectrum of Figure 20, the output peak powers at the two wavelengths of transmitted BP and Brillouin Stokes are 8.9 dBm and 9.8 dBm, respectively.

The impact of BP power on the peak power of Brillouin Stokes and the transmitted residual BP generated by the single-pass and dual-pass BFL is shown in Figure 21. The BP wavelength is set at around 1559.99 nm, while the BP power is varied from 12.2 dBm to 15.7 dBm. Below BP power of 12.2 dBm, the experiment is not continued as the SBS gain cannot sufficiently compensate for the loss inside the laser cavity and thus cannot support dual-wavelength generation. The peak power of generated Stokes is observed to increase as the BP increases and this can be attributed to the increase of the nonlinear Brillouin gain as the BP power arises. As shown in Figure 21, Brillouin Stokes line of single-pass configuration cannot meet the threshold requirement.

However, due to the stronger oscillation of the BP signal through nonlinear gain media, Brillouin Stokes of the dual-pass architecture can satisfy threshold power of about 14.5 dBm and reach to the saturation condition. Figure 22 also depicts the 14 times repeated scanning spectra of the two-wavelength lasing oscillations with a time interval of 10 minutes, which indicates the stability of the laser is very well. The maximum peak wavelength shifts at the two wavelength were both less than 0.002 nm measured by the OSA. The maximum power fluctuations at the peak wavelength of 1559.99 nm and 1560.07 nm were below 0.5 dB, which indicate an excellent stability in the dual-wavelength oscillation. The small power fluctuation is mainly due to the fluctuation of the input BP signal and a small quantity of Rayleigh components in the two lines which are unstable. Perhaps more stable dual-wavelength lasing output can be achieved if we try to improve the stability of the input BP and restrain or reduce the Rayleigh scattering. By tuning the BP wavelength, the wavelength of the Brillouin-stokes line is also changed correspondingly. Therefore, tunable dual-wavelength lasing output with a certain wavelength spacing of 0.08 nm can be achieved by adjusting the input BP wavelength.

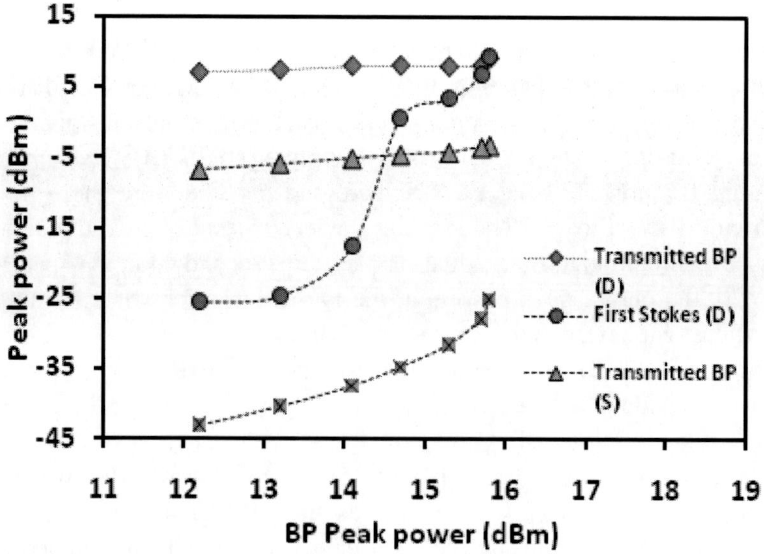

Figure 21. Residual BP and Stokes wave powers plotted against BP power.

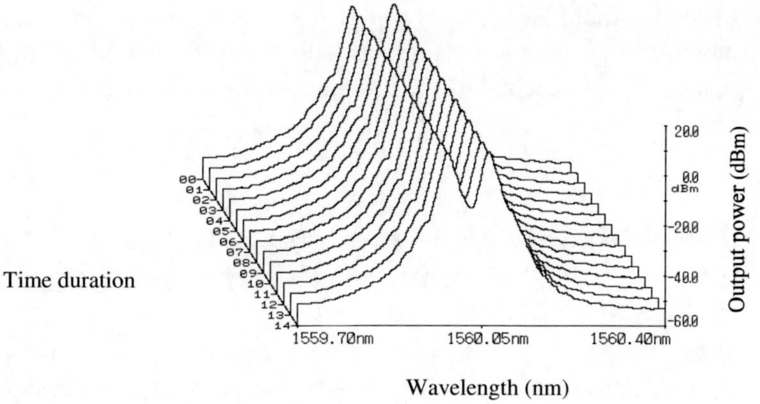

Figure 22. the spectral evolutions of the proposed BFL against time, which is repeatedly scanned every 10 minutes.

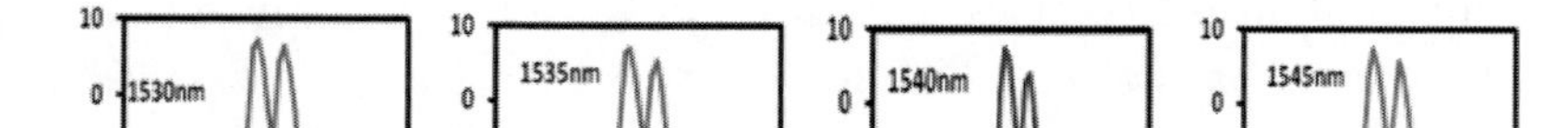

Figure 23. Residual BP and Stokes wave powers plotted against the input signal BP power.

Figure 23 shows the tunable dual-wavelength lasing spectra at the same BP power (~15.7dBm). As seen in this figure, amplitude-equilibrium dual-wavelength lasing outputs could be obtained in a tunable wavelength range of at least 35 nm from 1530 nm to 1565 nm. The tuning range is limited by the viability of broadband FBG and BP power. This is one of the remarkable advantages of Brillouin laser which can provide multi wavelength lasing in a broadband region of tuning wavelength as long as the BP with the fixed power in any wavelength is launched to the setup and the FBG's reflection region is used.

8. 0.16 nm Spaced Multi-Wavelength Brillouin Fiber Laser with Figure-of-Eight Configuration

To date, SBS-based multi-wavelength lasers have been investigated intensively [19, 33] due to their many advantages such as low threshold and narrow line-width, which are suitable for applications in microwave photonics [34], fiber sensors [30] and high capacity wavelength division multiplexing (WDM) communication networks [20]. Normally, a hybrid Brillouin/Erbium or Brillouin/Raman fiber lasers are used to achieve a multiple Stokes lines with a constant spacing and a high number of Stokes lines as reported in many literatures. For instance, the generation of 160 Brillouin Stokes has been experimentally demonstrated using a linear cavity self-seeded Brillouin/Erbium fiber laser (BEFL) [35]. Stable Brillouin/Raman fiber laser has also been demonstrated with 30 lasing lines by [36, 37]. However, the tuna-ability of these multi-wavelength lasers is limited by the Erbium or Raman gain bandwidths. The use of a Brillouin fiber laser (BFL) for multi-wavelength generation has not been considered seriously so far because of the limited number of lasing lines and the undesirable power flatness. In this section, we propose a new compact multi-wavelength BFL using a short piece of PCF as a nonlinear Brillouin gain medium in conjunction with a figure-of-eight configuration.

The PCF is used to significantly reduce the length of Brillouin gain medium._In this work, only 100m long PCF is used as a gain medium instead of a few km single-mode fiber in a conventional BFL. The proposed BFL can achieve stable multi-wavelength simultaneous lasing with 20 GHz or 0.16 nm line spacing, less threshold power and also capable to mitigate the limitation of tuning range. Compared to the previous BFLs and hybrid Brillouin lasers, the line spacing of the proposed BFL is twice as wide due to the removal of odd-order Stokes waves with the proposed figure-of-eight configuration and therefore it has more potential to be implemented in future WDM systems.

8.1. Experimental Configuration of Figure-Of-Eight

The configuration of the proposed figure-of-eight PCF-based multi wavelength BFL is shown in Figure 24. It includes an optical circulator, two couplers, and a 100 m long PCF as the Brillouin gain medium. The BP is sent into a piece of PCF from port 1 through port 2 of the optical circulator and 3 dB coupler (C1) to generate backward-propagating SBS. The backward-propagating SBS oscillates inside the first loop and the secondary loop in the clockwise (CW) and counter clockwise (CCW) directions respectively, to generate another Brillouin Stokes in the opposite direction (counter clockwise inside the first loop). The first Stokes signal is re-injected into the other end of the PCF in a clockwise direction to act as a BP to generate the next Stokes. The cascading process continues to generate multi-wavelength lasing. In the secondary loop, the odd-order Stokes lines are blocked by the circulator and the even-order Stokes lines are allowed to oscillate and build-up their intensity. The oscillating light inside the secondary loop is extracted from the laser system by using 90/10 coupler where 10% power is used for monitoring and measurement purpose while the 90% power is circulated back into the laser cavity.

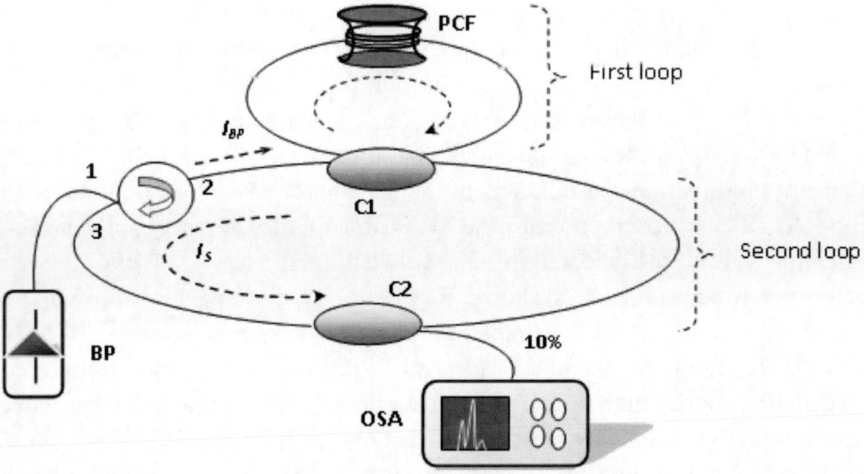

Figure 24. Configuration of the figure-of-eight PCF-based BFL.

The output spectrum is characterized by optical spectrum analyzers (OSA) with a resolution of 0.015 nm. The experiment is carried out for two different lengths of PCF; 50 m and 100 m. The experiments are also repeated with the conventional single ring cavity system, which is obtained by disconnecting the secondary loop and monitoring the output at port 3 of the optical circulator for comparison purpose.

8.2. Results and Discussion

Figure 25 shows the output spectra of the BFL with a single-ring cavity and a figure-of-eight configuration at two different lengths of PCFs; 50 m and 100 m. In the experiment, the BP power is fixed at 15.3 dBm. In the conventional set-up of single ring cavity, only the first Stokes travelling in the clockwise direction is obtained from the OSA as shown in Figure 25. Due to this fact, the single-ring cavity cannot provide sufficient nonlinear gain to assist in cascading process for multiple Stokes generation. It is found that the signal to noise ratios of 23.4 and 24.1 dB are achieved by the BFL with 50 m and 100 m PCF, respectively. In case of the figure-of-eight configuration, the BP generates the first Stokes, which has a higher peak power compared to the conventional configuration at wavelength shifted by 0.08nm from the BP as shown in Figure 25. The backward propagating Stokes generates the second Stokes which propagates in the opposite direction (counter clockwise direction in this setup). If the threshold condition is satisfied, this will generate a cascading Brillouin effect as a comb of lines. As shown in Figure 24, the oscillated signals in clockwise direction (odd order of Stokes lines) in the first cavity can be routed from C1 to port 3 of the circulator to oscillate in the secondary cavity in the counter clockwise direction. The circulator functions to ensure unidirectional oscillation in the secondary cavity. The output of the proposed BFL is observed from the 10% port of the output coupler and the remaining power is re-injected into the first cavity and act as a BP for the next even-order Stokes lines. As shown in Figure 25, the spacing between two consecutive odd-order Stokes lines is observed to be approximately 0.16 nm (~ 20 MHz in 1560 nm region). The anti-Stokes lines are also observed at wavelengths shorter than the BP, which occurs when the various Stokes waves interact with the pump wave through the four wave mixing (FWM) process. At the BP of 15.3 dBm and PCF length of 100 m, 7 lines are observed with the line spacing of 0.16 nm. Only 3 lines are observed with a shorter PCF length of 50 m.

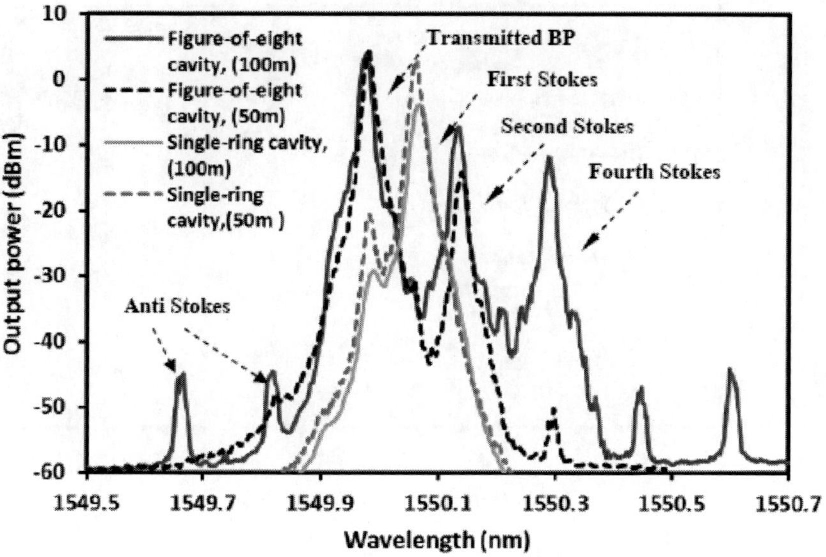

Figure 25. Output spectra of the BFL with single ring cavity and figure-of-eight configuration at PCF lengths of 50 m and 100 m. The BP power is fixed at 15.3 dBm.

Figure 26 shows the measured peak power for the Stokes at different BP powers for two different lengths of 50 and 100 m PCF in figure-of-eight MWBFL configuration.

The threshold power is defined as the BP power where the resonator starts to generate laser via SBS and below this threshold power, only spontaneous scattering emission is oscillating in the resonator. As shown in the figure, the peak power increases with the BP power. With PCF length of 100 m, the threshold powers for the 2nd, and 4th order Stokes lines are obtained at about 6.5 and 13.6 dBm respectively.

The 2nd Stokes of the conventional ring laser has a threshold power of larger than the maximum BP power of 15.3 dBm and therefore only the 1st Stokes is observed in Figure 25. However, with the addition of the secondary cavity in the proposed figure-of-eight cavity we managed to reduce the threshold pump power and thus generate additional Stokes lines without the need to add any linear or nonlinear gain media in the configuration. It was also found that the Stokes power begin to saturate as the BP power increases above a certain level. For instance, with 100 m long PCF, the second Stokes line saturates as the threshold for the fourth Stokes is reached.

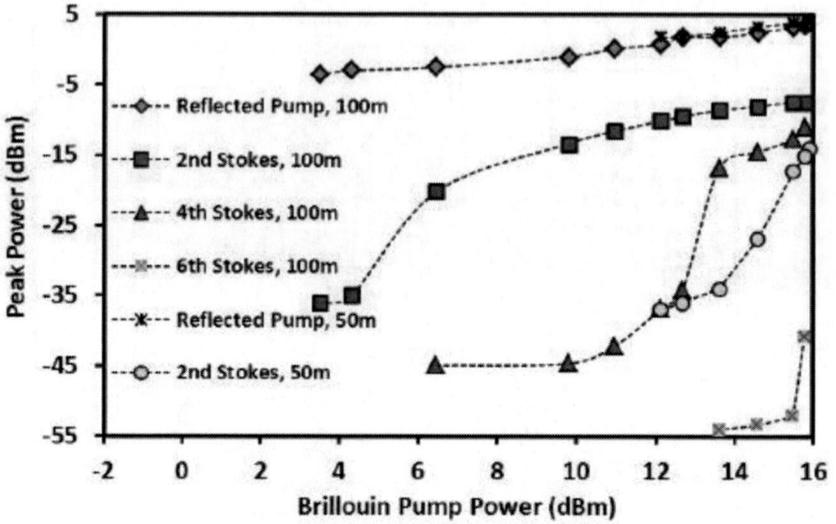

Figure 26. Brillouin Stokes power as a function of BP power for the proposed figure-of-eight BFL.

This is due to the power transfer from the second Stokes to the fourth Stokes, which occurs when the BP power is approximately at 12.1 dBm. The threshold power is also observed to be lower when a longer PCF is used. Figure 27 illustrates the output spectra of the proposed BFL as the BP wavelength is varied from 1520 nm to 1600 nm. It is found that the output spectrum is maintained over the 80 nm of tuning range. This is attributed to the multiple wavelength operation of the proposed BFL which is solely dependent on the SBS in the PCF but not much dependent on the operating wavelength.

As shown in the Figure 27, at least 3 even-order Brillouin Stokes lines are obtained at BP power of 12 dBm. This result shows that the proposed BFL has a remarkable advantage that it can provide multi-wavelength operation in a broad band tuning range as long as the BP with the fixed power can be launched into the setup. The stability of the multi-wavelength operation is also investigated since it plays a vital role for practical applications. Figure 28 depicts the spectral evolutions of the proposed figure-of-eightcavity multi-wavelength FBL against time. In the experiment, the output spectral is repeatedly scanned for every 20 min. The multi-wavelength BFL lases stably with power fluctuations of less than 0.5 dB for more than 3 hours. Temperature variations and mechanical vibrations were small as the experiment was conducted in a laboratory; thus the drift of the spectral profile was minimal. These results show that the proposed configuration has potentials for applications in the future WDM optical communication networks.

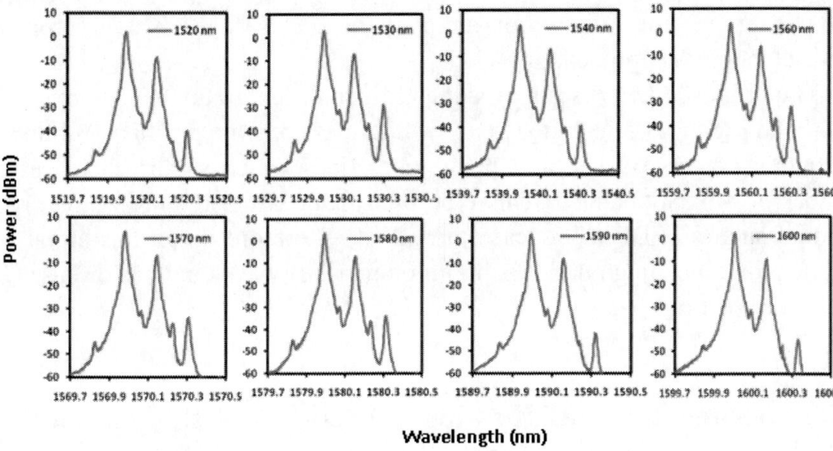

Figure 27. Output Spectra of the tunable MWFBL with 3 Brillouin Stokes lines with 0.16 nm line spacing and over 80 nm tuning range.

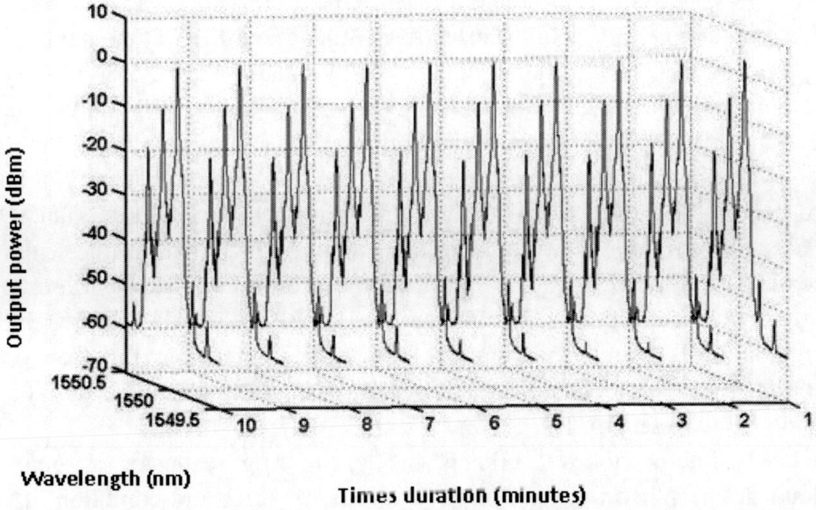

Figure 28. Spectral evolutions of the proposed figure-of-eight cavity multi-wavelength BFL against time. (The laser spectra scanned every 20 minutes).

9. MULTI-WAVELENGTH BRILLOUIN FIBER LASER WITH DUAL-PASS AMPLIFICATION CONFIGURATION

PCF-based BFLs have been demonstrated using various configurations for both single wavelength and multi-wavelength operations in the former sections [38]. In this section, a new configuration of a PCF-based BFL is proposed using a dual-pass approach.

The proposed BFL uses a broadband fiber Bragg grating (FBG) operating in C-band region as a reflector to allow dual-pass operation. This can improve the output powers of the cascading Stokes lines which in turn increases the number of Stokes obtained. This type of BFL does not require a linear gain medium such as EDF to increase the effect. Therefore it can operate at any wavelength depending on the Brillouin pump (BP) wavelength as discussed in the former sections.

9.1. Experimental Configuration of Dual-Pass Amplification

The proposed BFL architecture for the generation of multi-wavelength Stokes lines is depicted in Figure 29, with the inset showing the scanning electron micrograph (SEM) image of the PCF. It consists of a short piece of PCF, a 3dB coupler and a high-reflectivity broad band FBG at the cavity end. The proposed BFL configuration consists of two main sectors; a ring cavity as a Brillouin gain block which includes a piece of PCF and a 3 dB coupler, and a broad band FBG operating in C band region at the end of this linear structure to allow dual-pass Stokes lines oscillation in the ring resonator. The FBG has a 40nm bandwidth centered at 1545nm with a reflectivity of more than 99%. The BP is an external cavity tunable laser source (TLS) with a line-width of approximately 20MHz, which is amplified by an erbium-doped fiber amplifier to provide sufficient power for this study. The BP is launched into the PCF through a 3 dB coupler via optical circulator to generate the first order Brillouin Stokes in clockwise direction at wavelength downshifted by 0.08nm from the BP wavelength.

The Brillouin Stokes oscillates inside the ring resonator to generate Brillouin laser. When the laser power exceeds the threshold condition, it will then create the second order Stokes signal in the opposite direction. In the same way, the second order Stokes signal will create the third order Stokes signal.

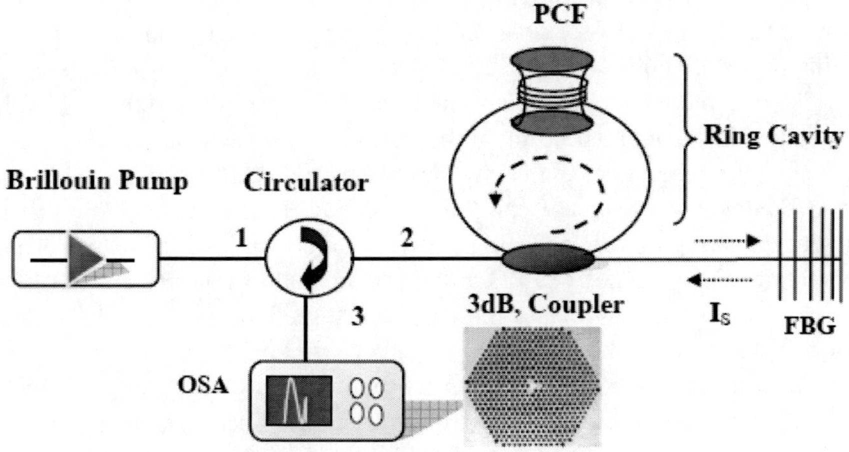

Figure 29. The proposed multi-wavelength BFL architecture.

The process will continue until the next higher order Stokes signals power is too small to exceed the threshold condition and hence creation of subsequent Stokes order signal will diminish. The residual BP and even order Stokes lines propagating in the clockwise direction are reflected by the broadband FBG to double propagating inside the ring resonator in the opposite direction. This will increase the nonlinear gain in the resonator which in turn improves the multi-wavelength lasing process. The output of the BFL is characterized using an optical spectrum analyzer (OSA) with a resolution of 0.015 nm.

CONCLUSION

Figure 30 compares the output spectrum of the BFL in different configurations at the maximum BP power of 15.7 dBm. First of all, the output of the BFL in the presence of the ring cavity but without FBG was studied. With this configuration, only the first Brillouin Stokes is achieved at 1560 nm wavelength with side mode suppression ratio (SMSR) of 26 dB. However, by incorporating the broad band FBG as in the proposed configuration of Figure 29, seven Stokes waves and four anti-Stokes waves with nearly 13 dB signal to noise ratio (SNR) are obtained by 100 m PCF around the oscillated BP at 1559.98 nm, whereas we have only the first Stokes wave in the same configuration by incorporating 50 m PCF. This is attributed to that the 50 m long PCF is not sufficient to create multi-wavelength in this configuration. By incorporating the 100 m long PCF, four anti-Stokes waves also arises due to four wave mixing (FWM) between co-propagating BP and BFL photons during the oscillation.

The bi-directional operation of the laser also contributes to the anti-Stokes generation. The output spectrum of the BFL is also investigated in the absence of ring cavity where the ring cavity is replaced by the PCF. In this BFL, only the first Stokes is observed with 2.5 dB SMSR. These results show that the presence of ring cavity and FBG is very important for multi-wavelength generation. The presence of FBG at the end side allows the BP and SBS light beams to propagate twice into the ring cavity section and thus the cavity is able to make more Stokes lines.

The impact of the BP power on the number of Stokes generated by the BFL is shown in Figure 31. The BP wavelength is set at around 1560 nm, while the BP power is varied from 6.1 dBm to 15.7 dBm. Below BP power of 6 dBm, the experiment is not continued as the SBS gain cannot sufficiently compensate for the loss inside the laser cavity and thus cannot support multi-wavelength generation.

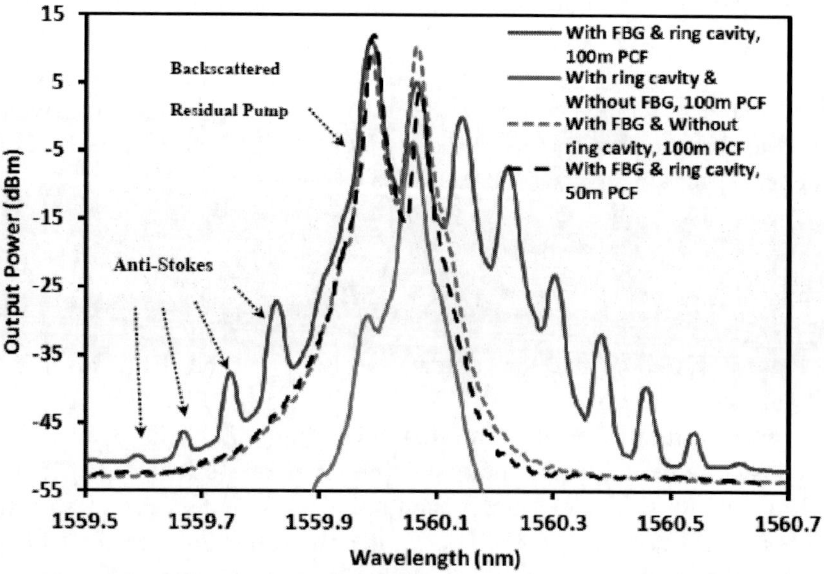

Figure 30. Output spectra of the BFL with different configurations and PCF's lengths.

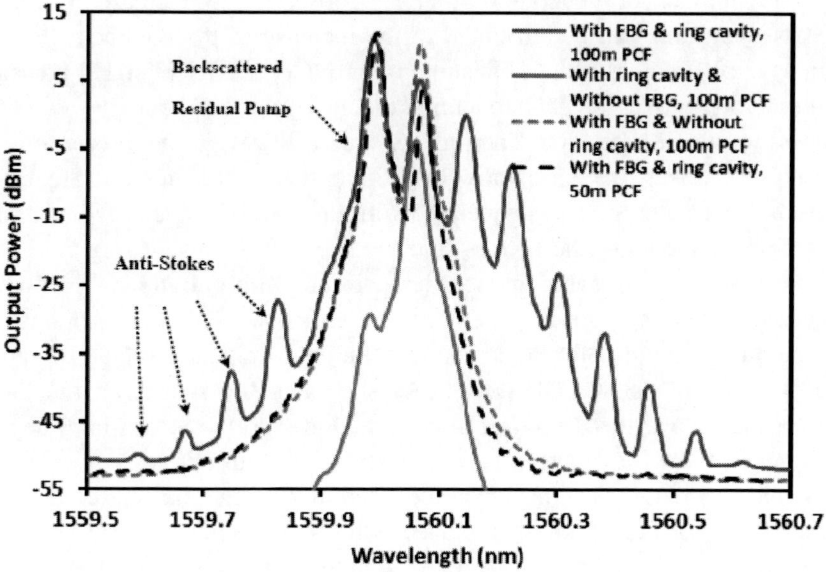

Figure 31. The output spectrum of the proposed BFL at various input BP power.

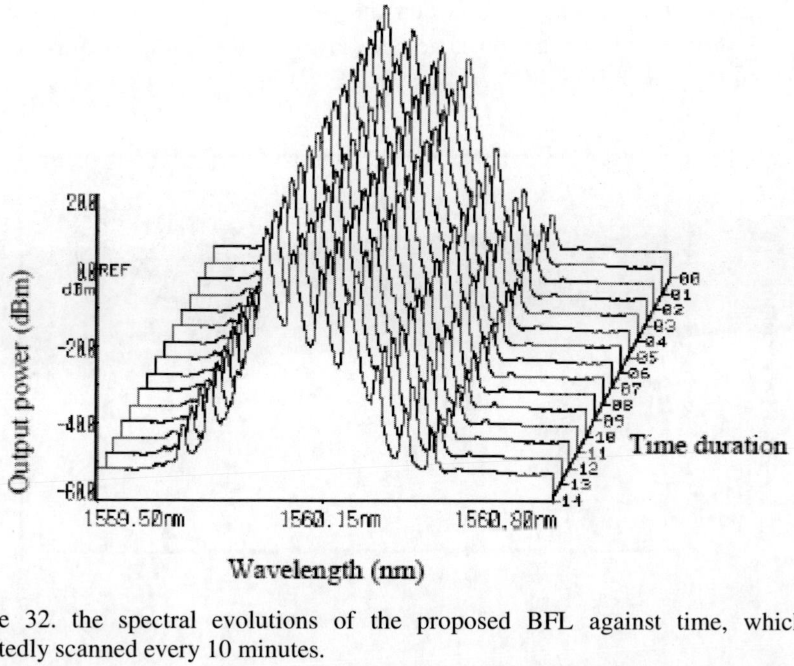

Figure 32. the spectral evolutions of the proposed BFL against time, which is repeatedly scanned every 10 minutes.

The number of generated Stokes is observed to increase as the BP increases and this can be attributed to the increase of the nonlinear Brillouin gain as the BP power rises. This situation provides sufficient signal power for higher order Stokes signals to pump the PCF and maintain the cascading process of the Stokes into multiple Stokes. However, the power of each subsequent Stokes line is typically lower than that of the previous Stokes line as each subsequent Stokes is generated with the energy of the previous Stokes, thus slightly reducing the Stokes line's power. As shown in Figure 32, the output of the BFL is stable at room temperature with only minor fluctuations observed coinciding with large temperature variances. Figure 33 demonstrates the output Brillouin Stokes power and the transmitted residual BP power against the input signal BP power. As shown in the figure, all the Stokes powers increase as the BP power increases. However, the Stokes lines begin to saturate as the BP power increases above the threshold power of the subsequent Stokes. For instance, the second Stokes line saturates as the threshold for the third Stokes is reached. This is due to the power transfer from the second Stokes to the third Stokes, which occurs when the BP power is approximately at 14.7 dBm. For the higher order Stokes, however, the threshold BP powers are obtained at almost the same level of 15.3 dBm. This is probably due to the FWM between the oscillating BP and Stokes signals which improves the conversion efficiency by providing amplification for the higher Stokes generation.

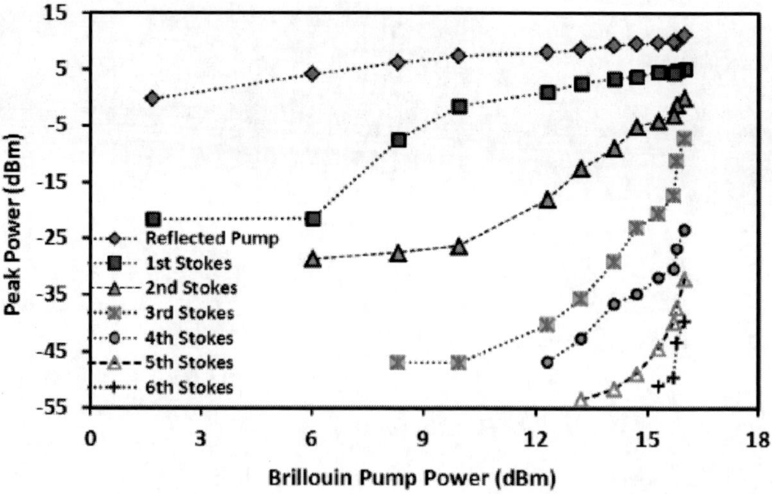

Figure 33. Residual BP and Stokes wave powers plotted against BP power.

REFERENCES

[1] Russell, P. St. J. (2006). Photonic-crystal fibers. *J. Lightwave Technol.*, 24: 4729-49.

[2] Laude, V., Khelif, A., Benchabane, S., Wilm, M., Sylvestre, T., Kibler, B., Mussot, A., Dudley J. M. and Maillotte, H. (2005). Phononic band-gap guidance of acoustic modes in photonic crystal fibers. *Phys. Rev. B.*, 71: 045107.

[3] Di Teodoro, F., Hoffman, P.R. (2005). Four-wave mixing self-stability based on photonic crystal fiber and its applications on erbium-doped fiber lasers. *Opt. Commun.*, 252: 111.

[4] Chremmos, I.D., Kakarantzas, G., Uzunoglu, N.K. (2005). Modeling of a highly nonlinear chalcogenide dual-core photonic crystal fiber coupler. *Opt. Commun.*, 251: 339.

[5] Geng, D., Beugnot, J.C., Sylvestre, T., Maillotte, H., M´elin, G., and Laude, V. (2007). Guided acoustic wave Brillouin scattering in photonics crystal fibers. *Opt. Lett.,* 2: 17–19.

[6] Beugnot, J.-C., Sylvestre, T., Alasia, D., Maillotte, H., Laude, V., Monteville, A., Provino, L., Traynor, N., Foaleng Mafang, S. and Th´evenaz, L. (2007). Complete experimental characterization of stimulated Brillouin scattering in photonic crystal fiber. *Optic Express*, 15(23):15517-22.

[7] Harun, S. W., Shahi, S. and Ahmad, H. (2009). Compact Brillouin–erbium fiber laser. *Opt. Lett.* 34: 46-48.

[8] Wu, Z. J., Shen, Q. S., Zhan, L., Liu, J. M., Yuan W. and Wang, Y. X. (2010). Optical Generation of Stable Microwave Signal Using a Dual-Wavelength Brillouin Fiber Laser. *IEEE Photon. Technol. Lett.*, 22(8): 568-70.

[9] Guan W. and Marciante J. R. (2010). Power scaling of single-frequency hybrid Brillouin/Ytterbium fiber lasers. *IEEE J. of Quantum Electronics,* 46(5): 674.

[10] Smith, R. G. (1972). Optical power handling capacity of low loss optical fibers as determined by stimulated Raman and Brillouin scattering. *Applied Optics,* 11: 2489-94.

[11] Shibata N., Okamoto K., and Azuma Y., (1989) Longitudinal acoustic modes and Brillouin-gain spectra for GeO2-doped-core single-mode fibers, *J. Opt. Soc. Am.* B, 6: 1167-1174.

[12] Azuma Y., Shibata N., Horiguchi T., and Tateda M., (1988) Wavelength dependence of Brillouin-gain spectra for single-mode optical fibers, *Electron. Lett.,* 24: 250-252.

[13] Nikles M, Thevenaz L, Robert P. (1996)Simple distributed fiber sensor based on Brillouin gain spectrum analysis. *Optics Letters,* 21(10): 758–760.

[14] Nobuyuki Y. and Imai T., (1993) Stimulated Brillouin scattering suppression by means of applying strain distribution to fiber with cabling, *Journal of Lightwave Technology,*11: 1519-1522.

[15] Pierre R. St., Mordaut D. W., Injeyan H., Bery J. G., Hilyard R. C., Weber M. E., Wickham M. G., Harpole G. M., Senn R., (1997) Diode array pumped kilowatt laser, *IEEE J. Sel. Topics Quant. Electron*. QE-3:. 53-58.

[16] Kuzin E. A., Petrov M. P., Davydenko B. E., (1985) Phase conjugation in an optical fibre, *Opt. Quant. Electron.* QE-17: pp. 393-397.

[17] Eichler H. J., Kunde J., Liu B., (1997) Quartz fibre phase conjugators with high fidelity and reflectivity, *Opt. Commun.* 139: 327-334.

[18] Harrison R. G., Kovalev V. I., Lu W., Yu D., (1999) SBS self-phase conjugation of CW Nd: YAG laser radiation in an optical fibre, *Opt. Commun.* 163: 208-211.

[19] Shirazi M. R., Harun S. W., Biglary M. and Ahmad H., (2008) Linear cavity Brillouin fiber laser with improved characteristics, *Optics Lett.,* 33(8): 770-772.

[20] Song Y. J., Zhan L., Hu S., Ye Q. H., and Xia Y. X., (2004) Tunable multiwavelength Brillouin-erbium fiber laser with a polarization-maintaining fiber Sagnac loop filter, *IEEE Photon. Technol. Lett.,* 16(9): 2015–2017.

[21] Agrawal G. P., (1989) *Nonlinear Fiber Optics*. Boston, MA: Academic,.

[22] Smith S. P., Zarinetchi F., and Ezekiel S., (1991) Narrow-linewidth stimulated Brillouin fiber laser and applications, Opt. Lett., 16: 393-395.

[23] Alahbabi M N, Cho Y T, Newson T P. (2005) 150-km-range distributed temperature sensor based on coherent detection of spontaneous Brillouin backscatter and in-line Raman amplification. *Journal of the Optical Society of America: B,* 22(6): 1321–1324.

[24] Nikles M, Thevenaz L, Robert P. (1996) Simple distributed fiber sensor based on Brillouin gain spectrum analysis. *Optics Letters,* 21(10): 758–760.

[25] Dai Y., Chen X., Sun J., Yao Y., Xie S., (2006) Dual-Wavelength DFB Fiber Laser Based on a Chirped Structure and the Equivalent Phase Shift Method, *IEEE Photon. Technol. Lett.,* 18(18): 1964.

[26] Li L., Schülzgen A., Zhu X., Moloney J.V., Albert J., Peyghambarian N., (2008) 1-W Tunable Dual-Wavelength Emission From Cascaded Distributed Feedback Fiber Lasers, *Appl. Phys. Lett.,* 92 (5): 051111.

[27] Sun J., Huang L., (2007) Photonic generation of frequency-switchable microwave signals exploiting polarization-induced spectrum splitting in fiber grating-based Fabry–Perot filters, *Opt. Commun.,* 273 (2): 482.

[28] Pan S., Zhao X., Lou C., (2008) Switchable single-longitudinal-mode dual-wavelength erbium-doped fiber ring laser incorporating a semiconductor optical amplifier, *Opt. Lett.* 33 (8) 764.

[29] Shahi , Harun S. W. and Ahmad H., (2009) Multi-wavelength Brillouin fiber laser using a holey fiber and a bismuth-oxide based erbium-doped fiber, *Laser Physics Letters,* 6(6), pp. 454-457.

[30] Han Y., Tran T.V.A., Kim S. and Lee S. B., (2005) Multiwavelength Raman-fiber-laser-based long-distance remote sensor for simultaneous measurement of strain and temperature, *Opt. Lett.* 30: 1282-1284.

[31] Harun S. W., Cheng X. S., Saat N. K. and Ahmad H., (2005) S-band Brillouin erbium fibre laser, *Electronics Letters*, 41(4): 174-176.

[32] Bjarklev A., Broeng J., and Bjarklev A. S., (2003*) Photonics Crystal Fibres,* Kluwer Academic Publishers.

[33] Gross M. C., Callahan P. T., Clark T. R., Novak D., Waterhouse R. B. and Dennis M. L., (2010)Tunable millimeter-wave frequency synthesis up to 100 GHz by dual-wavelength Brillouin fiber laser, *Optics Express,* 18(13): 13321-13330.

[34] Banchi L., Presi M., Proietti R. and Ciaramella E., (2010) System feasibility of using stimulated Brillouin scattering in self-coherent detection schemes, *Optics Express,* 18(12): 12702-12707.

[35] Zhan L., Li J. H., Xia J., Luo S. Y. and Xia Y. X., (2006) 160-line multiwavelength generation of linear-cavity self-seeded Brillouin-Erbium fiber laser, *Optics Express,* 14(22):10233-10238.

[36] Liu Y., and Wang D., (2008) Wavelength tunable and amplitude-equilibrium dual-wavelength lasing sources with dual-pass Raman/Brillouin amplification configuration *Optics Express*, 16(6): 3583-3588.

[37] Liu Y., Wang D. and Dong X., (2008) Stable room-temperature multi-wavelength lasing oscillations in a Brillouin–Raman fiber ring laser, *Optics Communications,* 281: 5400–5404.

[38] Yang, X., Dong X., Zhang S., Lu F., Zhou X., and Lu C., (2005) Multiwavelength erbium-doped fiber laser with 0.8-nm spacing using sampled Bragg grating and photonic crystal fiber, *IEEE Photonics Technol. Lett.*, 17: 2538–2540.

In: Photonic Crystals
Editor: Donald Claus
ISBN: 978-1-62618-193-9

Chapter 3

PHOTONIC CRYSTAL TUNABLE FILTER DRIVEN BY AN ELECTRICAL CONTROL SIGNAL

***A. Wirth L. Jr.*[1,2,*] *and A. S. B. Sombra*[1,2]**

[1]Laboratory of Telecommunications and Materials Science and Engineering, Fortaleza, Ceará, Brazil

[2]Department of Teleinformatics Engineering, DETI, Center of Technology, Federal University of Ceará (U.F.C.), Fortaleza, Ceará, Brazil

ABSTRACT

We are presenting a novel model of tunable filters based on photonic crystal waveguides and resonant cavities to work in high transmission rates optical systems.

The insertion of an electrical command signal in the dielectric rods, which function as diodes, increases the conductivity, but the optical loss is very small. On the other hand, the increase in the conductivity generates an increase in the refractive index. The tuning of the required optical channel is achieved by means of the proper voltage inserted in the diodes.

Keywords: Optical device, photonic crystal, tunable filter

---

* Corresponding author. E-mail: awljeng@gmail.com.

1. INTRODUCTION

Researchers around the world are increasingly using photonic crystal (PhC) because it enables the achievement of new optical devices, for example, waveguides [1], lasers [2], splitters [3], antennas [4], optical switches [5, 6], etcetera.

Due to the nanometer dimension of the periodicity found in photonic crystals, we can obtain micrometer devices, which can be integrated within photonic integrated circuits (PICs).

We designed and analyzed a tunable filter embedded in photonic crystal, which is controlled by an electrical command signal. This device may operate in optical systems with high transmission rates, thus contributing to the achievement of fully optical telecommunication systems.

The PhC structure is constituted by a silicon on insulator (SOI) slab with triangular lattice of dielectric rods embedded in a substrate (SiO2), whose refractive index is $n_s = 1.55$ at λ (wavelength) = 1.55052 μm.

The dielectric rods have radius $r_b = 0.2a$, where "a" is the lattice constant. Given that this device will operate in the C-band of the ITU, which covers the wavelength range from 1528.77 nm to 1560.61 nm [7], we adopted the refractive index of the silicon dielectric rods according to the Sellmeier equation, as we will see later.

The tunable PhC filter consists of a resonant cavity embedded between two waveguides (input waveguide and output waveguide) obtained by inserting a defect line in the middle of the device. We designed the resonant cavities based on our research and simulations.

To achieve the tuning of the filter, an electrical command signal is inserted into the doped dielectric rods (localized in the resonant cavity region), which work as diodes. An optical modulation in a p-i-n-diode-embedded photonic crystal (PhC) silicon Mach–Zehnder interferometer modulator was examined [8], however, we used a different process from the above presented, since we used doped rods in the resonant cavity region as direct polarized diodes, as we will see later. The voltage of the electrical command signal inserted in the doped rods (diodes) causes only a small increase of the rods conductivity in the coupling region and consequently an increase in the refractive index of these rods. Taking into account that the increase of the optical loss value is negligible, the effective refractive index increases causing a decrease in the value of the coupling coefficient. On the other hand, the desired change in the refractive index depends only on the voltage of the command signal.

This device is extremely fast and it can be integrated into PICs.

This paper is organized as follows: In Section 2, we present the resonant cavity design used in our device. Section 3 is devoted to the detailing of the electrical control and Section 4 to the tunable filter. We conclude this paper in Section 4 with our conclusions.

2. Resonant Cavities Design

We know that the resonant cavities quality factor values (Q) of the modes can be calculated by the following equation:

$$Q = 2\pi f_0 \frac{U}{P} = \frac{w_0 \tau}{2}, \quad (1)$$

where f_0 is the resonant frequency of the cavity mode, U is the time averaged energy stored in the cavity, P is the power loss and τ is the lifetime (fields in the cavity decays according to $e^{-t/\tau}$). Hence, Q is dimensionless, i.e., Q is the number of optical periods occurred until the energy in the resonant cavity decays to $e^{-2\pi}$.

Indeed, the Q value results from a number of possible factors, including intrinsic material absorption, etch-induced surface roughness and surface-state absorption, and other fabrication irregularities [9].

If we consider a resonant cavity located in the middle of a waveguide, the vertical radiation produces light emission. On the other hand, the in-plane radiation can result in transmission of a signal inserted into one end of the waveguide (input waveguide) through the resonant cavity, which leaves the device through the other waveguide (output waveguide).

It is worth noting that the minimizing of the vertical radiation (in this case is considered as optical loss) is obtained by the maximizing of the vertical quality factor ($Q_\perp$). Hence, we need to achieve resonant cavities with $Q_\perp > 10^5$ for certain devices, such as narrow bandwidth filters [10]. However, the in-plane quality factor ($Q_{||}$) value is governed by the in-plane confinement of the defect mode, which depends on the Distributed Bragg Reflection (DBR) of the surrounding photonic lattice. Therefore, the in-plane confinement of the defect mode can be determined by the number of periods of the PhC lattice surrounding the defect, as well as the angular frequency and the in-plane guided mode bandgap.

3D photonic crystals can control light in the three dimensions, but it is very difficult to fabricate these devices at optical wavelengths. As a general rule the planar photonic crystals, also known as photonic crystal slabs, have been used to allow an easier fabrication. Planar photonic crystals combine 2D photonic crystals confinement of the light with a slab waveguide in the vertical direction, i.e., index confinement is used to confine the light in the third dimension.

Although the so-called "air-bridge" slab structure consists of a thin 2D photonic crystal in a high-index membrane surrounded by air, this type of structure is not easily integrable into a chip. Hence, we used a hybrid structure often found in SOI-based PhC, where the lower cladding consists of a structured oxide layer, whose effective refractive index is close to 1 [11], so that the device with this structure is easier to be integrated into a chip than a membrane and is almost symmetrical.

In a purely 2D PhC, the Transverse Magnetic (TM) modes (odd modes) have no electric field in the horizontal middle plane (x-y plane) of the PhC and have magnetic field in the x-y plane. Moreover, the TM modes have an electrical field in the z direction, and no magnetic field in the z direction. TE modes (even modes) present the reverse of what occurs in TM modes. However, the finite height of the photonic crystal slabs leads polarization mixing and the modes are not purely TM (or TE)-polarized anymore. On the other hand, if the (x, y)-plane in the middle of the slab is a mirror plane of the structure, the first-order modes are very similar to the corresponding modes existing in infinite 2D photonic crystals.

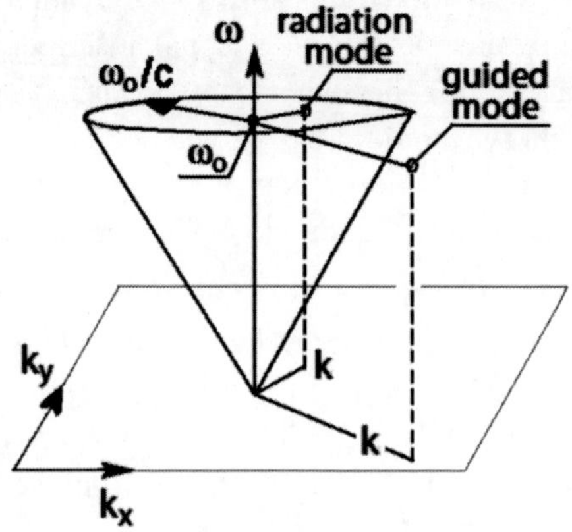

Figure 1. Light cone of an air-clad PhC cavity.

Furthermore, in the (x, y)-mirror plane itself, these modes are purely TM (resp. TE) polarized. In other words, in this case the polarization mixing is quite small and the approximation TE for even mode (TE-like) and TM for odd modes (TM-like) can be assumed.

While the effective vertical wavelength of the TE-like modes is more dependent on the high dielectric material, the effective vertical wavelength of the TM-like modes is more dependent on the low dielectric material. For this reason, the gap for hole slabs opens up at a smaller thickness (h) than for rod slabs.

The light of a PhC slab cone consists of radiation modes that are extended infinitely in the region outside the slab, where the boundary between guided and radiation modes, is called "light line".

In a PhC slab consisting of a uniform material above and below the slab core, the light line is constituted by the modulus of the wave vectors divided by the cladding refractive index and the cladding refractive index must be lower than the averaged index of the slab core in order to confine the light in the vertical direction.

In Figure 1 we show the light cone referring to an air-clad PhC slab. The vertical confinement for an air clad PhC slab cavity depends on the vertical radiation loss, which occurs when the wave-number (k) lies within the light cone. On the other hand, we may note that to have a guided mode, the wave-number (k) must be outside the light cone. Therefore, if we insert a mode with wave-number (k) outside the light cone at the input of a waveguide, for example, this signal goes out the device through the waveguide output.

We can define the transmission spectrum as the ratio between the input signal power and the output signal power as a function of the angular frequency.

By means of the coupled-mode theory, it is possible to find the total quality factor (Q) and the transmission spectrum as a function of Q as shown below [10]:

$$Q = \left(\frac{1}{Q_{\|}} + \frac{1}{Q_{\perp}} \right)^{-1} \tag{2}$$

$$T(w) = \frac{\frac{1}{4Q_{||}^{2}}}{(w-w_0)^2 + \frac{1}{4Q^2}}, \tag{3}$$

where w_0 is the resonant angular frequency and w are the possible values of the angular frequency within the bandwidth. Equation 3 represents a Lorentzian peak, whose maximal value is located at $w = w_0$. In this case the maximum peak transmission is:

$$T(w_0) = \left(\frac{Q}{Q_{||}}\right)^2 = \left(\frac{Q}{Q_{||}+Q_{\perp}}\right)^2 \approx 1 - \frac{2Q}{Q_{\perp}} + Q\left(\frac{Q_{||}}{Q_{\perp}}\right)^2. \tag{4}$$

From equation 4 we can conclude that for $Q_{\perp} >> Q_{||}$ ($Q_{\perp} > 10^5$ as it was previously mentioned) the transmission peak is approximately 100%, because the cavity mode decays much more quickly into the waveguide than into the air. Therefore, we need to find resonant cavity structures for modes with wave-number (k) outside of the light cone with appropriate $Q_{\perp}$ and $Q_{||}$ and values, which in turn enable greater performance of the device due to the minimization of the radiation loss and the maximization of the in-plane transmission spectrum.

When we excite the cavity mode directly in the middle of the resonant cavity with high $Q_{||}$ value, then the energy in the resonant cavity escapes slowly and it penetrates in the waveguides. However, when we insert an optical signal within one of the waveguides (input waveguide) with a well determined value of $Q_{||}$ the transmitted power to the other waveguide (output waveguide) can be high, depending on the frequency of the input signal (w). In this case, the quality factor of the cavity is given by $w_0/\Delta w$, where Δw is the full width at half-maximum of the transmission spectrum (Δw_{FWHM}).

Taking into account the rules of the G.692 regulation of the International Telecommunication Union – T (ITU-T) [7], we can assume that the spacing between two adjacent channels is 50 GHz ($\Delta f_{FWHM} = 50Ghz$) (or $\Delta f_{FWHM} = 100Ghz$). On the other hand, the required time for the incident pulse to be transmitted through the cavity determines the optical switching time, which is dependent on the spectral width of the resonant cavity. Therefore, we need that the quality factor of our resonant cavity enables a high transmission rate and hence the lowest possible switching time.

With $Q = f_0/\Delta f_{FWHM}$, we obtain the value of the ideal quality factor of our resonant cavity: 3870 for $\Delta f_{FWHM} = 50$ Ghz and 1930 for $\Delta f_{FWHM} = 100$ Ghz (spectral widths of the signals, according to the ITU-T).

Taking into account that the structure of our PhC resonant cavities is composed of silicon dielectric rods in silica, we needed to take several simulations in order to obtain the cavities whose parameters are the closest possible of the ideal PhC parameters.

We choose the best value of the lattice constant and thickness (706 nm) of our SOI rod slab, based on the value of the gap size (gap/midgap ratio) and on the light cone and used the plane wave expansion (PWE) method to find the PhC dispersion relation and photonic bandgap (PBG) for z-odd modes (between z-odd_band1 and z-odd_band2).

In Figure 2 is showed the dispersion relation referring to the PhC slab with $h = 706$ nm and $r = 0.2a$. The normalized frequency is defined as $u = a/\lambda$, where "a" is the lattice constant and λ is the wavelength.

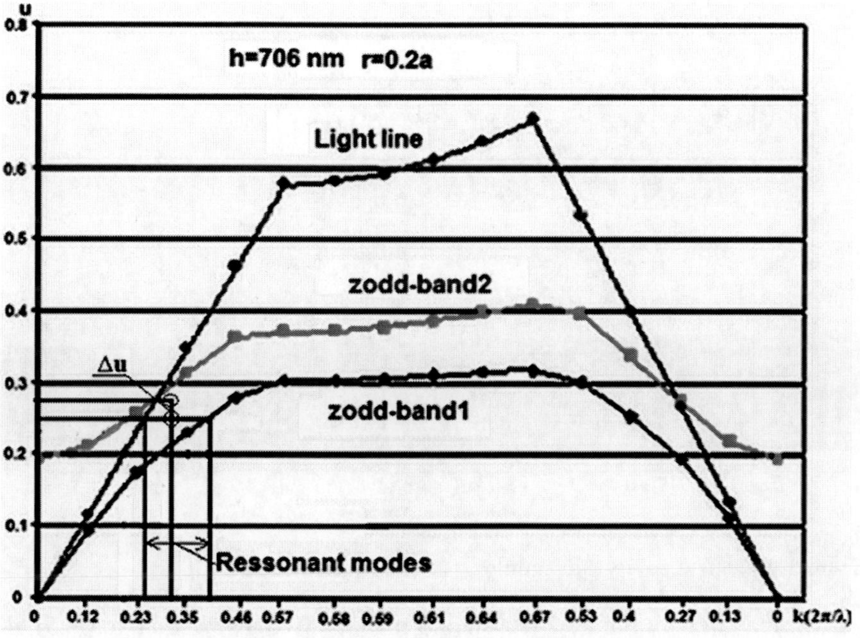

Figure 2. Dispersion relation of the photonic crystal slab.

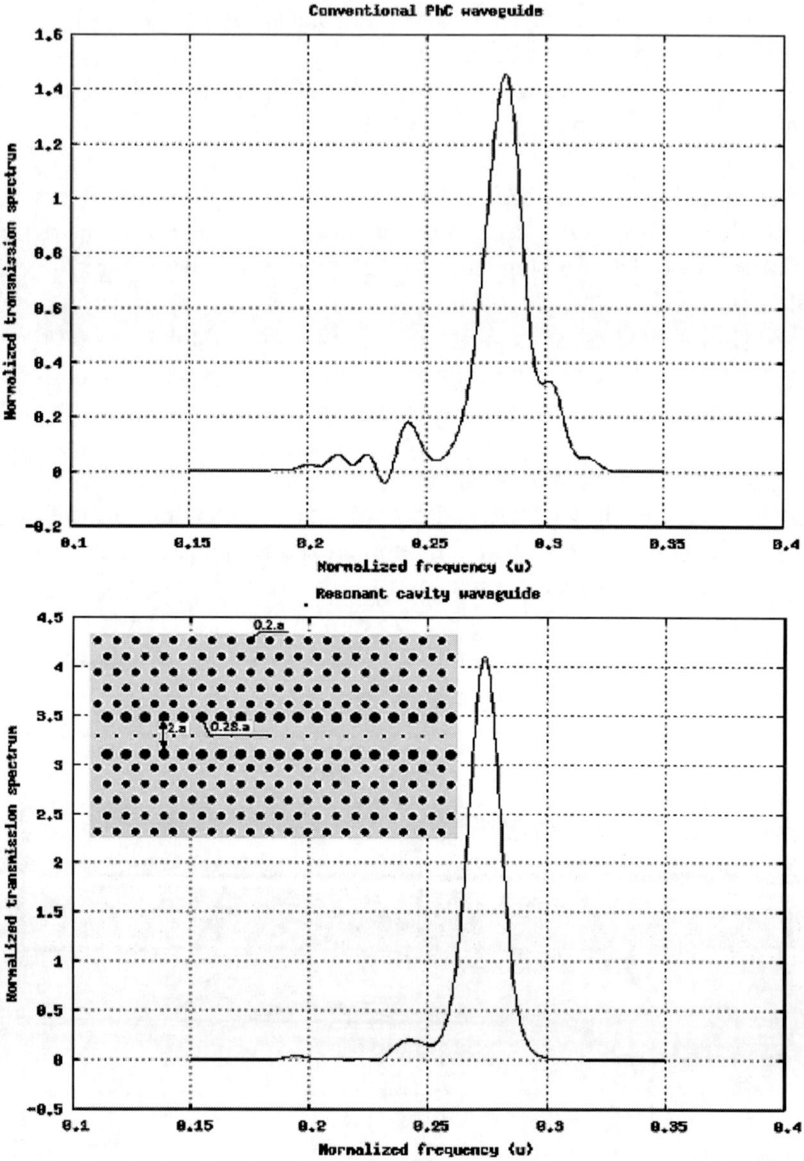

Figure 3. Resonant cavity waveguide.

We may note in Figure 2 (for the normalized frequency equal to 0.25), that the resonant modes localized outside the light cone must have the value of the wave-number vectors (k) between *0.25 (2π/a) and 0.383 (2π/a),* which are located in the light cone and in the first band, respectively. On the other hand, as we are interested in finding a filter to operate according to the rules of the ITU, we choose $\lambda = 1.55$ μm so that for $u = 0.25$ we must have $a = 388$ nm.

Taking into account that in the linear regime $k = n_{eff}.(2\pi/\lambda)$ (n_{eff} is the effective refractive index of the mode) and $k = x.(2\pi/a)$, where x is the abscissa in Figure 2 ($0,25 < x < 0.383;\ u = 0.25$), we obtained: $1 < n_{eff} < 1.53$. On the other hand, if the mode is located outside the light cone, we have that $x > u$, so that $x.(2\pi/a) > (a/\lambda).(2\pi/a)$, i.e, $k > k_0$. Following a similar reasoning, we get $x = u$ and $k = k_0$ if the mode is located in the light cone. Similarly, $x < u$ and $k < k_0$ if the mode is located inside the light cone. The above detailed confirms what was said with respect to Figure 1.

We designed and analyzed four resonant cavities by means of the finite-difference time-domain (FDTD) method working around *you* = *0.25* as detailed below, where the computational cell is surrounded by a perfectly matched layer (PML) medium.

The waveguides embedded in the PhC structure (which together with the resonant cavities form the tunable filter) are obtained by removing one row of dielectric rods. Taking into account that the transmission spectrum regarding these waveguides must be improved, we performed several simulations by means of FDTD and reached the conclusion that the width of the waveguides need be increased from $a.3^{1/2}$ to $2.a$. Moreover, dielectric rods were inserted with radius equal to $0.08.a$ (or $0.09.a$ as we will see later) within the waveguide and the radii of the dielectric rods of the first rows above and below the central region of the waveguide were changed from $0.2.a$ to $0.28.a$. At the top of the Figure 3 we can see we can see the graph of the normalized frequency ($|E^2|$, where E is the electric mode amplitude) versus normalized frequency (u). The peak amplitude of the pulses is also normalized, in order to compare the transmission spectrum of the structures.

At the bottom of the Figure 3 we can find the best transmittance performance of the used waveguide (resonant cavity waveguide) in relation to the conventional PhC waveguide. The input pulse covers the whole normalized frequency (u) range of the PhC. In the inset of Figure 3 we show the obtained waveguide structure.

Although the performance of the above detailed waveguide is very good, we obtained a waveguide with even better performance by changing the radii of the dielectric rods of the first rows above and below the central region of the waveguide from $0.28 \times a$ to $0.32 \times a$ and the radii of the five dielectric rods located in the second rows above and below the central region of the waveguide from $0.2 \times a$ to $0.28 \times a$, as it can be observed by

Figure 4. We can notice that there is a compression in the transmitted pulse within this optimized waveguide and an increase in the transmission.

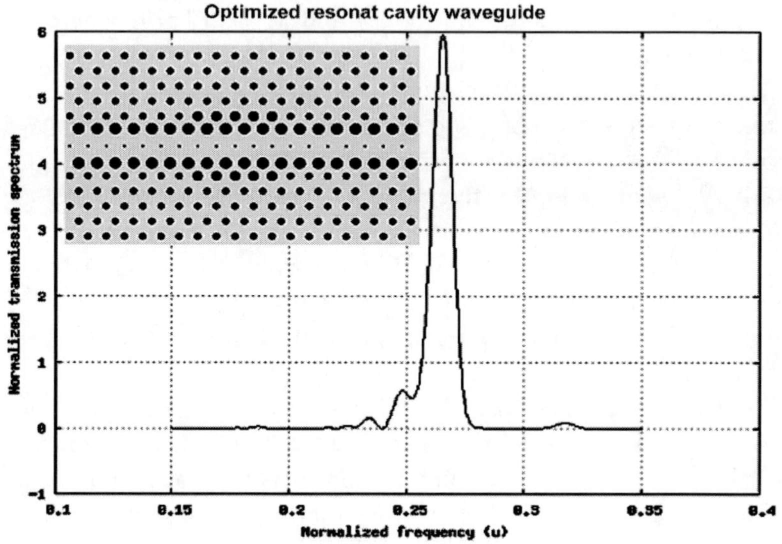

Figure 4. Optimized resonant cavity waveguide.

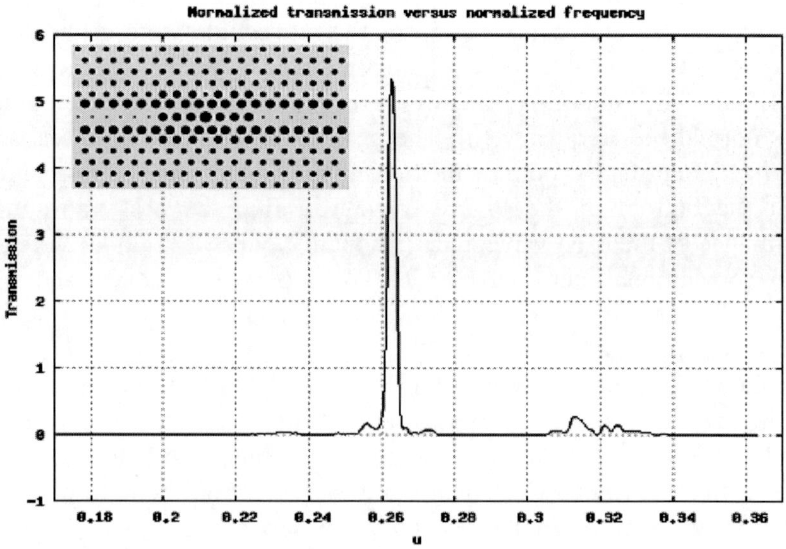

Figure 5. Resonant cavity 1.

The first resonant cavity (cavity 1) was obtained by inserting a dielectric rod with a radius $r = 0.4 \times a$ in the center of the waveguide, and the insertion of three dielectric rods with radius $r = 0.31 \times a$ beside the central dielectric rod. Moreover, we changed the radius of the dielectric rods in the second rows above and below the central dielectric rod located in the waveguide center, as it can be seen from Figure 5 (obtained by the FDTD method).

It is worth noting, that the greater the number of dielectric rods located beside the central dielectric rod (N), the higher the quality factor (Q). However this quality factor saturates close to the value of the perpendicular quality factor (Q_p). Hence, as Q_p practically does not change, we can calculate the value of Q_p according to the above detailed ($N \geq 6$) inserting an optical pulse in the center of the resonant cavity. The Q_p value found in resonant cavity 1 was only *1,178* (ideally $Q_p > 10^5$). Hence, this resonant cavity suffers a great optical loss (radiation loss). The total quality factor regarding the resonant cavity frequency ($u = 0.263$) is $Q = 244$.

The second cavity is shown in Figure 6. Previous studies in PhCs with hexagonal structures of dielectric perforated with air holes showed, that the elongation of the holes along the symmetry axes can be used to obtain cavities with high quality factor and small modal volume [12]. However, this PhC structure is quite different from that used in our device. Taking into account that the used resonant cavities PhC structure are formed by silicon rods embedded in silica we modified the central cylindrical dielectric rod by an elliptical dielectric rod and designed the structure in a different way to that presented by [12] (Figure 6).

We define the spacing (e) between the ellipse and the dielectric rods located in the central region of the waveguide as follows: $e = 0.5 - (r_3 \times 2)$, where r_3 is the radius of the rods dielectric located beside the ellipse. Consequently, the ellipse in the (x, y)-plane has length equal to $l_x = 2 \times (1 - r_3 - e)$ in the x-axis direction. As the length of the ellipse in the y-axis direction does not contribute significantly in transmittance, we adopted this length equal to $0.8 \times a$. After several simulations, we found the values of r_3 and l_x that provide the largest transmittance ($r_3 = 0.17 \times a$; $l_x = 1.34 \times a$). Moreover, the remaining dielectric rods inserted within the waveguide have a radius equal to $0.09 \times a$ and the radii of the dielectric rods located in the second rows above and below the central region of the waveguide were changed to $0.32 \times a$, with the exception of the radii of the dielectric rods in the center of the second rows, whose value were changed to $0.29 \times a$.

The Q_p value for this cavity is *26,060* and the Q value of the resonant frequency ($u = 0.259$) is *121*. Hence, this resonant cavity suffers moderate radiation loss. The total quality factor regarding this resonant cavity frequency (*u* = *0.263*) is *Q* = *121*.

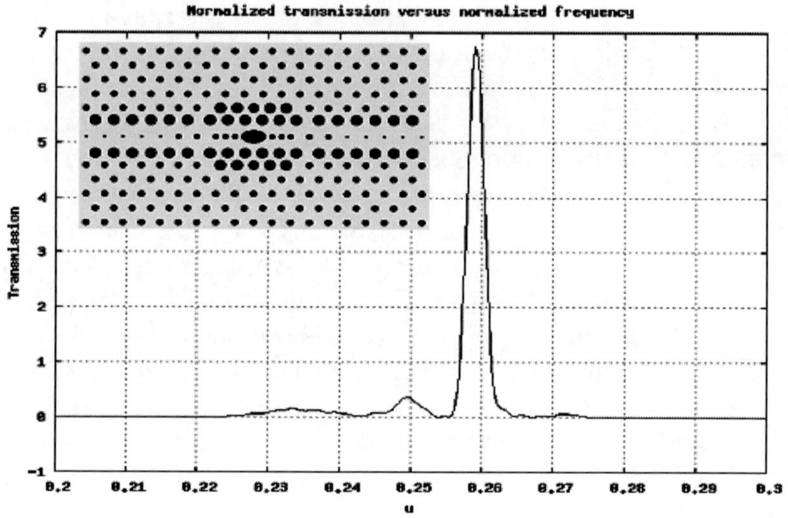

Figure 6. Resonant cavity 2.

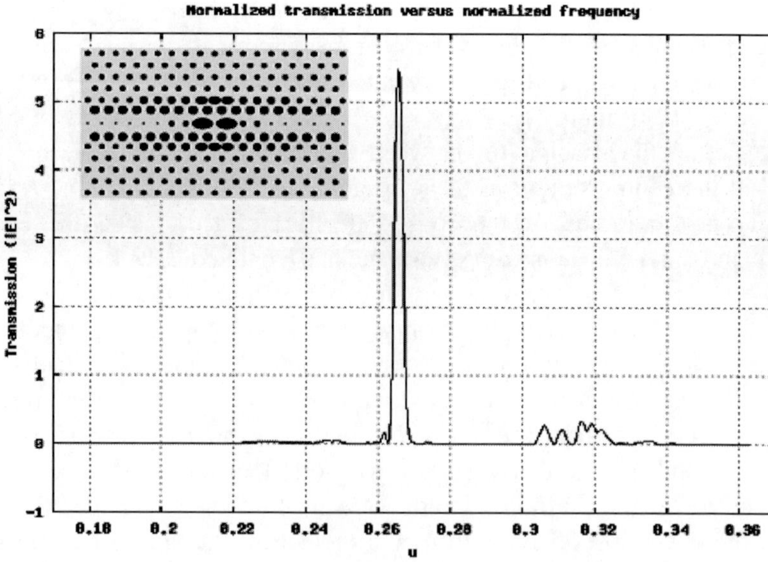

Figure 7. Resonant cavity 3.

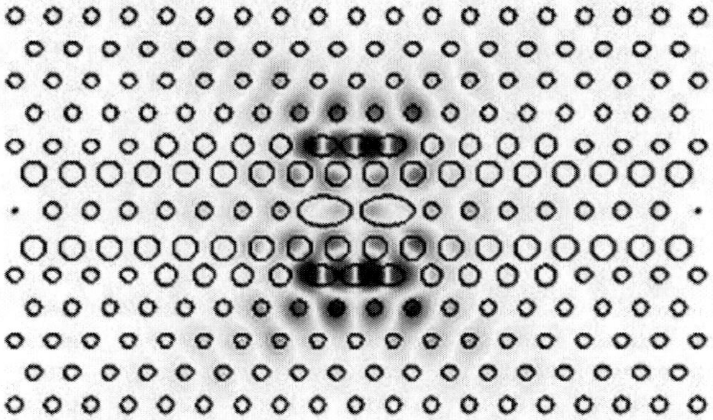

Figure 8. Field distribution pattern of the resonant mode in cavity 3.

The resonant cavity 3 is presented in Figure 7.

After several simulations, we found the values of r_3 and l_x that provide the largest transmittance ($r_3 = 0.21 \times a$; $l_x = 1.44 \times a$). Moreover, the remaining dielectric rods inserted within the waveguide have a radius equal to *0.08 x a* and the radii of the dielectric rods located in the second rows above and below the central region of the waveguide were changed to *0.28 x a* but we used three elliptical dielectric rods in the center of the second rows ($lx = 3 \times r_3$; $Li = 2x\, r_3$), as we can see in Figure 7.

The Q_p value for this cavity is 129,617. We can see the field distribution pattern of the resonant mode (N = 7) by Figure 8. It is worth noting that the radiation loss is quasi-null, so that we can neglect the radiation loss.

The total quality factor regarding this resonant cavity frequency ($u = 0.263$) is $Q = 145$.

3. ELECTRIC CONTROL DETAILING

The insertion of an electrical command signal in the dielectric rods of the resonant cavity, which function as diodes, increases the conductivity, but the optical loss is very small. The increase in the conductivity generates an increase in the refractive index. The required increase of the conductivity and consequently the required increase in the refractive index are obtained from the value of the voltage inserted in the diodes.

Several articles were published, which detail the implementation of Diodes, LEDs, Lasers and Modulators with nanometer dimensions, embedded in photonic crystals. An optical modulation in a p-i-n-diode-embedded photonic crystal (PhC) silicon Mach–Zehnder interferometer modulator was examined [13]. Electrical operation of III-Nitride light emitting diodes (LEDs) with photonic crystal structures was demonstrated [14]. Light emitting pn-diodes were fabricated on a 5.8 μm thick n-type Si device layer of a silicon-on-insulator (SOI) wafer using standard silicon technology and boron implantation. The thickness of the Si device layer was reduced to 1.3 μm, corresponding to a 4λ-cavity for λ=1150 nm light [15]. However, we used a different method from those presented above, since we used the rods in the coupling region of the couplers as direct polarized diodes.

In doped semiconductors, the typical impurity concentration is 10^{22} donors (N_D) or acceptors (N_A) atoms/m^3. This generates concentrations 5-6 orders of magnitude higher than the intrinsic silicon. Therefore, the number of free carriers is very close to the doping level. Doping of a semiconductor reduces the concentration of minority carriers. In this case, the conductivity is caused mainly by the impurity atoms:

$$\sigma = qn\mu_n \text{ (N-Type crystals)}, \tag{6}$$

where q is the electron charge, n_q is the number of free electrons ($\approx N_D$) and μ_n is the electron mobility in silicon.

$$\sigma = qp\mu_n \text{ (P-Type crystals)}, \tag{7}$$

where q is the electron charge, p is the number of holes ($\approx N_A$) and μ_p is the hole mobility in silicon.

The rods of the coupling region are doped, so that each rod of the coupling region works like a PN-diode, which are connected to a low voltage source.

We can use equations (6) and (7) to obtain the minimum conductivity (without voltage): 2.928 (Ω. cm)$^{-1}$. The depletion zone width is given by:

$$w = \sqrt{\frac{2\varepsilon_s(Vb - V)}{q}\left(\frac{1}{N_A^-} + \frac{1}{N_D^+}\right)}, \tag{8}$$

where V_b is the voltage of the potential barrier of the depletion zone, ε_S is the Silicon electric permittivity and V is the external electric voltage. Thus, the length of the rods should be calculated to allow the maximum depletion zone width and as well as the areas for *carrier* recombination. In Figure 9 we show the graph of the depletion zone width due to the direct voltage applied to the terminals of the diode.

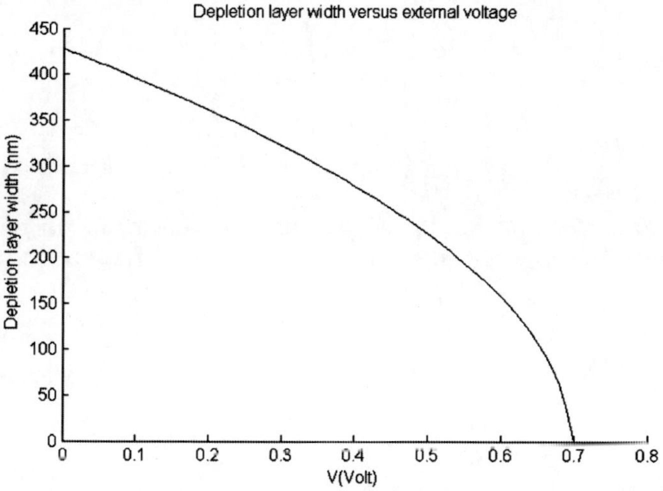

Figure 9. Depletion zone width versus direct voltage.

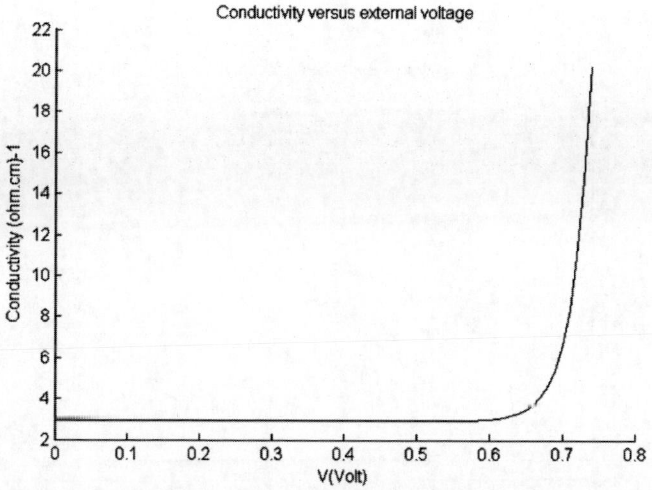

Figure 10. Conductivity ($\Omega^{-1}.cm^{-1}$) versus direct voltage.

We know that the excess of carrier at the depletion region edges, i.e., $-X_p$ (p-side) and X_n (n-side) are given by:

$$\Delta(n_p) = (-X_p) = n_{p0}\left(e^{qV/kT} - 1\right) \tag{9}$$

$$\Delta(p_n) = (X_n) = p_{n0}\left(e^{qV/kT} - 1\right), \tag{10}$$

where $n_{p0} = \frac{n_{i2}}{NA}$, $p_{n0} = \frac{n_{i2}}{ND}$ are the minority carrier density, n_i is the number of free electrons and holes per cubic centimeter in silicon, V is the applied external voltage, k is the Boltzmann constant = $1.38*10^{-23}$ J/°K and T is temperature in degrees (Kelvin).

When $V > V_b$ appears a current through the diode. We know that the current caused by the presence of a direct external electric voltage applied to a PN junction diode is given by:

$$I = I_s\left(e^{qV/kT} - 1\right) \tag{11}$$

where I_s is the reverse saturation current of minority carriers for the silicon (10^{-12} A).

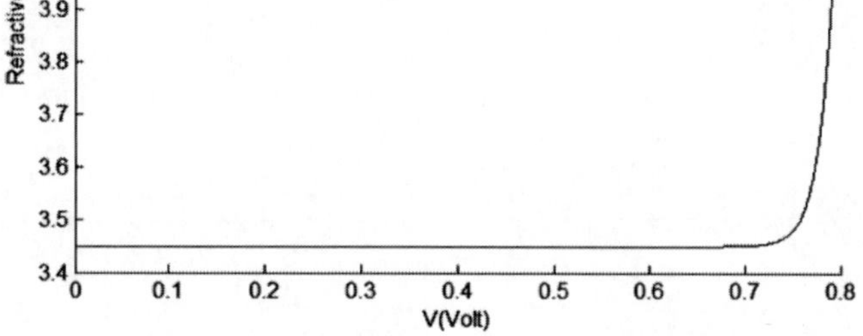

Figure 11. Magnitude of the refractive index due to the direct voltage.

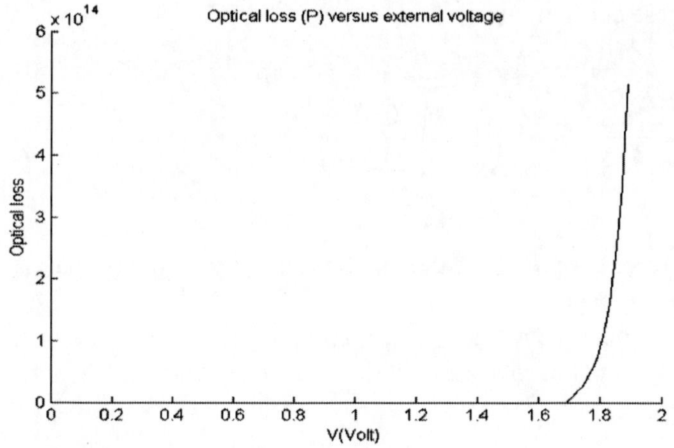

Figure 12. Optical loss (Np/m) as a function of the direct voltage.

Based on Equations 9, 10 and 11, we show the graph of conductivity as a function of the applied direct voltage (Figure 10).

From the solution of the wave equation in the absence of magnetic effects, we obtain the complex refractive index given by:

$$n = \sqrt{\varepsilon_r - i\frac{\sigma}{w\varepsilon_r\varepsilon_0}} = \sqrt{\varepsilon_r - i\frac{\sigma\lambda}{2\pi c\varepsilon_r\varepsilon_0}}, \qquad (12)$$

where ε_r is the dielectric constant of the medium.

In Figure 11 we show the graphic of the refractive index magnitude for silicon as a function of *V*, obtained from Equation 12.

For example, if we want to obtain n = 3.65 the voltage must be V = 0.7775V.

However, this value is theoretical and very high relative to values obtained so far.

From the solution of the wave equation, we have that the attenuation constant is obtained by:

$$\alpha = \frac{2\pi c}{\lambda}\sqrt{\frac{\mu\varepsilon}{2}\left[\sqrt{1+\left(\frac{\sigma\lambda}{2\pi c\varepsilon}\right)^2} - 1\right]} \text{ (Np/m)} \qquad (13)$$

and the phase constant by:

$$\beta = \frac{2\pi c}{\lambda}\sqrt{\frac{\mu\varepsilon}{2}\left[\sqrt{1+\left(\frac{\sigma\lambda}{2\pi c\varepsilon}\right)^2}+1\right]} \quad \text{(rad/m)} \qquad (14)$$

In Figure 12 is presented the graph of the optical loss (Np/m) as a function of the direct voltage.

We may note from Figure 12, that the optical losses due to the imposed conductivity can be neglected until the voltage equal to 1.5V.

4. Tunable Filter

Equation 15 is the Sellmeier equation for the relationship between the silicon refractive index and the mode wavelength.

$$n^2 = 1 + \frac{10.6684293\lambda^2}{\lambda^2 - 0.301516485^2} + \frac{0.003043475\lambda^2}{\lambda^2 - 1.13475115^2} + \frac{1.54133408\lambda^2}{\lambda^2 - 11104^2} \qquad (15)$$

We used the FDTD method to find the normalized frequency of the cavity resonant mode adopting the mode refractive index according to the Equation 15, so that cavity 1 has a resonant normalized frequency $u_{r1} = 0.2634$. For the filtering of an optical channel with normalized frequency $u_{11} = 0.2601$, we need to change the material refractive index of the dielectric rods ($\Delta n = 0.05$). If we want to filter another optical channel, for example, with normalized frequency $u_{12} = 0.2566$, we need to use $\Delta n = 0.1$.

In Figure 13 we show the transmittance referring to the three channels as function of Δn in the tunable filter based on cavity 1.

Cavity 2 has a resonant normalized frequency $u_{r2} = 0.259$. For the filtering of an optical channel with normalized frequency $u_{21} = 0.259$, the change in the material refractive index of the dielectric rods must be $\Delta n = 0.07$. If we want to filter another optical channel, for example, with normalized frequency $u_{22} = 0.2546$, we need to use $\Delta n = 0.14$.

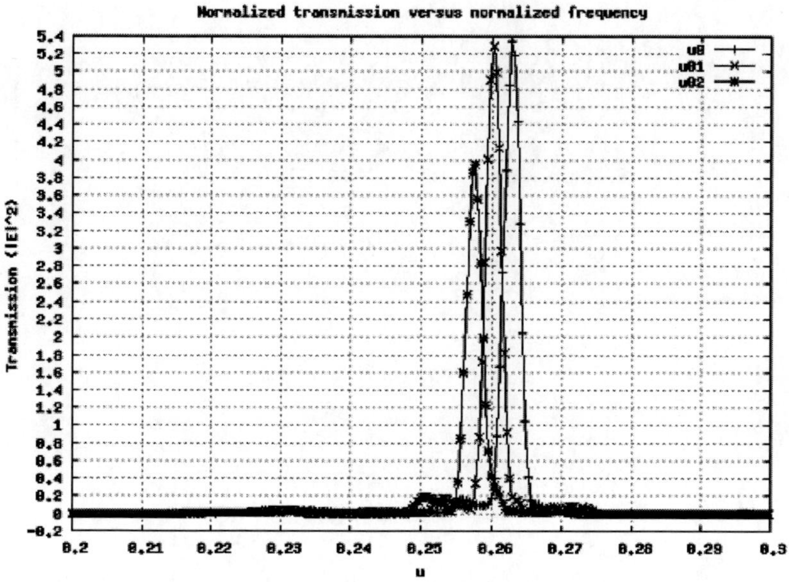

Figure 13. Transmittance in the tunable filter (based on cavity 1).

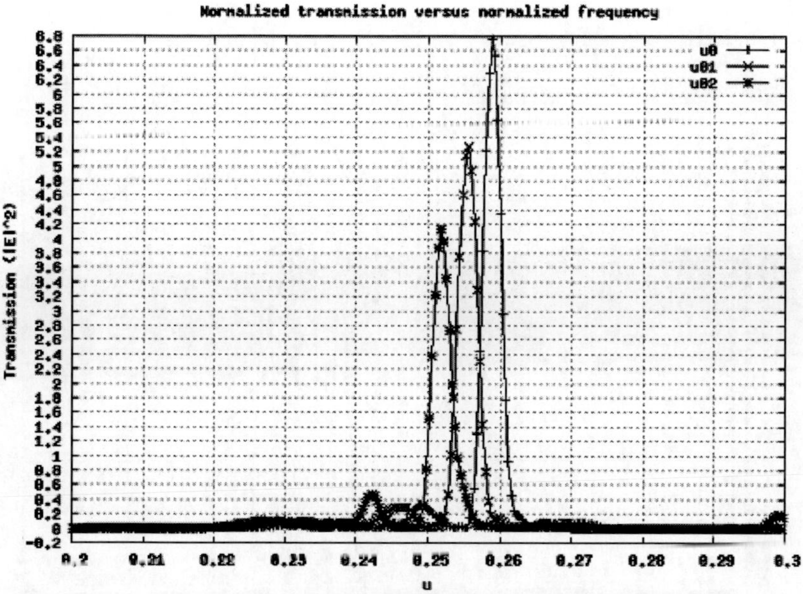

Figure 14. Transmittance in the tunable filter (based on cavity 2).

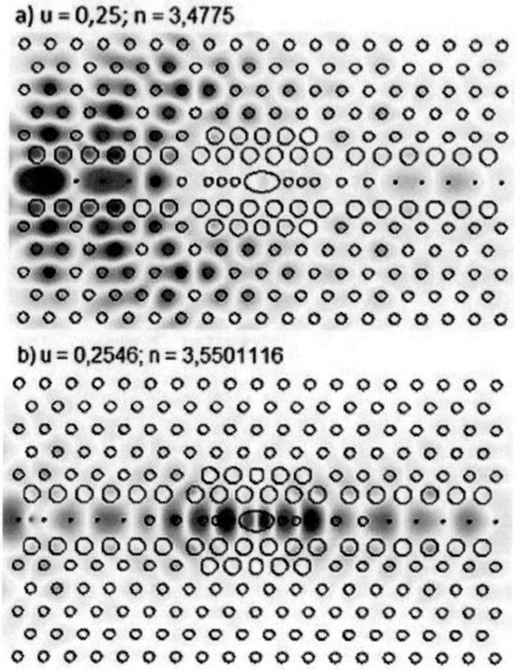

Figure 15a. Reflected channel (Cavity 2). 15b. Filtered channel (Cavity 2).

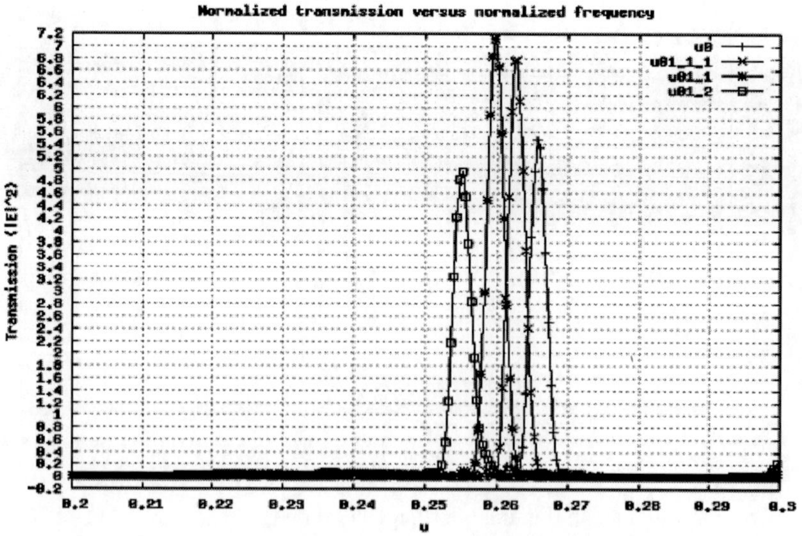

Figure 16. Transmittance in the tunable filter (based on cavity 3).

In Figure 14 we show the transmittance referring to the three channels as function of Δn in the tunable filter based on cavity 2.

We use the FDTD method to show by Figure 15a that the channel with normalized frequency u = 0.25 and material refractive index n = 3.4775 ($\Delta n = 0$) is totally reflected (not filtered). On the other hand, the channel with u = 0.2546 and n = 3.55 ($\Delta n = 0.14$) is not reflected, so it is filtered (Figure 15b).

In fact, by increasing the material refractive index, the resonant frequency shifts to higher frequencies. In Figure 16 is shown the shift of the resonance frequency in relation to the resonance frequency obtained without changing the material refractive index on the cavity 3 (u_0 = 0.2656). Applying the change in the material refractive index equal to 0.055, the frequency is shifted to u_1 = 0.262, so that the channel spacing of the filtered modes is determined by modifying the material refractive index value.

In Figure 16 we can see four filtered channels (u_0 = 0.2656; n = 3.4775, $u_{01\_1\_1}$ = 0.2629; n = 3.5328, u_{01_1} = 0.2596; n = 3.5822 and u_{01_2} = 0.2553; n = 3.6601).

Although there are materials that undergo material refractive index reduction [16], is very hard to get the decrease and increase of the resonant frequency, because this procedure requires the increase and decrease, respectively, of the material refractive index value of the same device. Therefore, we get only the decrease of the resonant frequency by increasing the material refractive index.

CONCLUSION

In summary, we have investigated a tunable filter embedded in a photonic crystal. We designed three resonant cavities embedded in a PhC structure, which consists of silicon dielectric rods embedded in silica. The tuning of the optical channels is obtained by changing the material refractive index due to the insertion of an electrical command signal in the dielectric rods of the cavities, which work as diodes.

Therefore, we are presenting a tunable filter based on the waveguide resonant cavity to work in optical telecommunication networks.

ACKNOWLEDGMENTS

This work was partly sponsored by Coordination for the Improvement of Higher Level -or Education- Personnel (CAPES).

REFERENCES

[1] M. Loncar, D. Nedeljkovic, T. Doll, J. Vuckovic, A. Scherer, and T. P. Pearsall, Waveguiding in planar photonic crystals, *Appl. Phys. Lett.,* vol. 77, pp. 1937–1939, (2000).

[2] Adil Karim, Student Member, Joachim Piprek, Patrick Abraham, Dan Lofgreen, Yi-Jen Chiu and John E. Bowers, 1.55-_m Vertical-Cavity Laser Arrays for Wavelength-Division Multiplexing, *Quantum Electronics*, Vol. 7, No. 2, pp. 178 – 183, (2001).

[3] R.W. Ziolkowski and M. Tanaka, FDTD analysis of PBG waveguides, power splitters and switches, *Opt. Quant. Electron.*, vol. 31, pp. 843–855, (1999).

[4] B. Temelkuran, M. Bayindir, E. Ozbay, R. Biswas, M. M. Sigalas, G. Tuttle, and K. M. Ho, Photonic crystal based resonant antenna with a very high directivity, *J. Appl. Phys.*, vol. 87, pp. 603–605, (2000).

[5] P. R. Villeneuve, D. S. Abrams, S. Fan, and J. D. Joannopoulos, Single-mode waveguide microcavity for fast optical switching, *Opt. Lett.,* vol. 21, pp. 2017–2019, (1996).

[6] A. Wirth L. Jr., M. G. Silva, A. C. Ferreira and A. S. B. Sombra, "All-optical nonlinear switching cell made of photonic crystal", *J. Opt. Soc. Am. A*, Vol. 26, No. 7, pp. 1661 – 1667, (2009).

[7] International Telecommunication Union (ITU- T, G.692), Series G: Transmission Systems and Media Digital Systems and Networks, (10/98).

[8] Lanlan Gu, Wei Jiang, Xiaonan Chen, and Ray T. Chen, "Physical Mechanism of p-i-n-Diode-Based Photonic Crystal Silicon Electrooptic Modulators for Gigahertz", *IEEE Journal of Selected Topics in Quantum Electronics*, Vol. 14, No. 4, pp. 1 – 7, (2008).

[9] K. Srinivasan and O. Painter, Momentum space design of high-Q photonic crystal optical cavities, *Opt. Express* 10, 670 (2002).

[10] J. D. Joannoupoulus, S. G. Johnson, J. N. Winn and R. D. Meade, "Photonic Crystals – Moldind the Flow of Light", Priceton University Press, Second edition, (2008).

[11] C. Jamois, R.B. Wehrspohn, L.C. Andreani, C. Hermann, O. Hess, U. Gösele, "Siliconbased two-dimensional photonic crystal waveguides", Elsevier, Photonics and Nanostructures – Fundamentals and Applications 1 pp. 1 -13, doi:10.1016/ j.photonics.2003.10.001, (2003).

[12] J. Vuckovic, M. Loncar, H. Mabuchi and A. Scherer, Optimization of the Q factor in photonic crystal microcavities, *IEEE J. of Quantum Electron.* 38, 850 (2002).

[13] Lanlan Gu, Wei Jiang, Xiaonan Chen, and Ray T. Chen, "Physical Mechanism of p-i-n-Diode-Based Photonic Crystal Silicon Electrooptic Modulators for Gigahertz", *IEEE Journal of Selected Topics in Quantum Electronics*, Vol. 14, No. 4, pp. 1 – 7, (2008).

[14] J. J. Wierer et all, "III-Nitride LEDs with photonic crystal structures", *Proc. of International Society for Optics and Photonics – SPIE*, Vol. 5739, pp. 102 – 107, (2005).

[15] J. Potfajova, J.M. Sun, B. Schmidt, T. Dekorsy, W. Skorupa and M. Helm, "Silicon-on-insulator microcavity light emitting diodes with two Si/SiO2 Bragg reflectors", *Journal of Luminescence*, Volume 121, Issue 2, pp. 290-292, (2006).

[16] Brian R. Bennett, Richard A. Soref, and Jesus A. del Alamo, Carrier-Induced Change in Refractive GaAs, and InGaAsP, *IEEE J. of Quantum Electronics*, Vol. 26, No. 1, pp. 113 - 122, (1990).

In: Photonic Crystals
Editor: Donald Claus
ISBN: 978-1-62618-193-9

Chapter 4

REVIEW OF RECENT PATENTS ON TUNABLE PHOTONIC CRYSTALS

Anirudh Banerjee*
Department of Electronics and Communication,
Amity School of Engineering and Technology, Lucknow,
Amity University, Uttar Pradesh, India

ABSTRACT

This chapter reviews some patents on tunable Photonic Crystals for use in different application areas. These selected patents find potential applications in optical communications and optical sensing. The chapter also suggests an idea which can be explored to further optimize the tunability of photonic band gaps in photonic crystals.

Keywords: Photonic crystals, external stimulus, tunable spectra

1. INTRODUCTION

Photonic crystals (PCs) are periodically structured electromagnetic media, possessing photonic band gaps [1-4] i.e. ranges of frequency in which light can

* Email: anirudhelectronics@yahoo.com.

not propagate through the structure. A 3D photonic crystal is shown in figure 1.

Figure 1. A 3D photonic crystal.

The periodicity of the structure is of the order of the wavelength of light. The photonic band gaps are analogous to the band gaps or energy gaps for electrons traveling in semiconductors. In the case of semiconductors, a band gap arises from the wave-like nature of electrons. Electrons as waves within a semiconductor experience periodic potential from each atom and are reflected by the atoms. Under certain conditions, electrons with certain wavevectors and energy constitute standing waves. The range of energy, named "bandgap" in which electrons are not allowed to exist, appears. In a similar manner, standing waves of electromagnetic waves can be formed within a periodic structure whose period is of the order of the light wavelength. In this case, the medium expels photons with certain wavelengths and wavevectors. Such a structure acts as an "insulator of light", and this phenomenon is referred to as photonic band gap formation.

Recently, photonic crystals (PCs) have attracted great interest of researchers due to their remarkably useful capabilities to guide and control light propagation. In these photonic band gap structures, generally, two or more, different optical materials are periodically arranged. These periodic multilayer structures lead to the formation of photonic band gaps or stop bands, in which propagation of electromagnetic waves of certain wavelengths are prohibited. These bands or ranges depend upon a number of parameters such as refractive indices of materials, thicknesses of material layers and angle of incidence etc. [4]. If all these parameters are kept constant, then this PC structure will have fixed predetermined bands of wavelengths or frequencies which will be reflected and transmitted by the structure. By changing one or more of these parameters and keeping other parameters constant in the structure, wavelengths transmitted through the structure can be tuned or changed. Actually, the transmission and reflection wavelength bands obtained in the output spectra of these PCs shift when one or more of these parameters are changed. Therefore, these PC structures can act as tunable optical filters [4-5] and optical sensors [6-7]. Several designs of tunable PCs have been suggested and fabricated by researchers [8-14]. In these tunable PCs, tuning of transmission and reflection bands were achieved through variation of one or more of these structural parameters electrically, electro-optically, thermally, mechanically, chemically, magnetically etc. For optical sensing, by monitoring the transmission or reflection band change or shift, a change in structural parameters can be detected. Based on this, refractometric [6], temperature [7], mechanical stress sensing [11] PC devices etc. have been proposed and fabricated by researchers. In this chapter, some recent patents on tunable photonic crystal structures for application in different areas are reviewed. This tunability feature of PCs makes them very dynamic, multipurpose and useful.

2. Selected Recent Patents on Tunable Photonic Crystals

In patent [8], the tunable photonic crystal comprised of electrically active and elastic polymer network in which refractive indices of the structural materials was tunable through an external electrical and/or mechanical stimulus. The controllable change of refractive indices in the photonic crystal structure by using electrically-conductive and mechanically-deformable component caused the reflectance wavelength shift in response to the external stimulus. The method of manufacturing tunable photonic crystal particles comprised of forming a template, infiltrating the template with a polymer precursor mixture, curing the polymer precursor mixture to form a polymer network, removing the template from the polymer network; and comminuting the polymer network to form particles. Thus, a photonic crystal was formed with a reflectance wavelength.

In patent [9], an infrared emitter, utilized a photonic crystal (PC) structure having semiconductor material layer, a dielectric material layer overlaying the semiconductor material layer, and a metallic material layer having an inner side overlaying the dielectric material layer. The semiconductor material layer was capable of being coupled to an energy source for introducing energy to the semiconductor material layer. An array of surface features was defined in the device in a periodic/quasi-periodic manner. The emitter device was adaptable to emit electromagnetic energy having spectral characteristics determined by parameters of the periodically distributed surface features, the parameters including shape, size, depth, distribution geometry, periodicity, material properties and defects. Thus, this photonic crystal could produce tunable electromagnetic emissions with a narrow band of wavelengths.

The invention [10] was a tunable photonic crystal based drop filter apparatus capable of extracting and redirecting any selected at least one wavelength from a source signal including multiple wavelengths. The photonic crystal drop filter has a photonic crystal having first waveguide for transmitting light having a frequency within a band gap of the photonic crystal, and a second waveguide. The resonant cavity modified the transmission characteristics of the first waveguide by creating one or more transmission zeros that comprise narrow frequency ranges within the bandgap of the photonic crystal material at which light that is otherwise capable of being transmitted through the first waveguide is prevented from propagating through the first waveguide. The second waveguide was connected to the first waveguide by a resonant cavity for redirecting and extracting the wavelength of the light prevented from propagation through the first waveguide. A moving device was connected to the dielectric tuning member to move it to the desired position in the second waveguide to allow selection of wavelengths to be extracted. Actually, the wavelength of light that can be extracted from the first waveguide is a function of the position of the dielectric tuning member with respect to the resonant chamber. The apparatus is particularly suitable as an extraction device for WDM optical communications systems where it is required to extract one or more carrier signals from a set of carrier signals.

In patent [11], a flexible photonic crystal device, comprising of a periodic array of high index dielectric material was embedded in a flexible polymer. A dynamic and real time tunability was achieved by the application of a variable force with a MEMS actuator. The force induces changes in the structural parameters of the photonic crystal, and consequently modifies the photonic band structure. The effect of mechanical stress on the anomalous refraction behavior of the flexible PC was significant, and a very wide tunability in beam propagation direction was observed. For real time applications high quality flexible PC structures were fabricated by e-beam lithography and anisotropic etching processes.

In application [12], a first photonic crystal structure has a first photonic band gap distribution and is configured to support a first electromagnetic signal, this first photonic crystal structure includes a first thermally responsive material which is configured to vary dimensionally in response to a second electromagnetic signal different from the first electromagnetic signal. Therefore, the first photonic band gap distribution is configured to vary correspondingly to the dimensional variation of the first thermally responsive material.

In patent [13], a tunable photonic crystal device comprises of a working electrode and a counter electrode forming a cell space with a conductive solution filling the cell space. The conductive solution comprises charged particles and a solvent with a wavelength tunable photonic crystal film in the cell space. The material comprises an ordered array of first constituents having a first refractive index embedded within a polymer network having a second refractive index other than the first refractive index. The ordered array of first constituents has a lattice spacing giving rise to a reflected wavelength. The polymer network being expandable and contractible in response to controlled uptake and expulsion, respectively, of at least one of the charged particles and the solvent by the polymer network. The reflected wavelength depends on the lattice spacing and the lattice spacing is tunable by expansion and contraction of the polymer network. The amount of uptake and expulsion of at least one of the charged particles and the solvent is tunable by controlling an electronic configuration of the polymer network.

In patent [14], there is provided a tunable photonic crystal device comprising alternate layers of first and second materials. The alternate layers comprise of a responsive material. The responsive material is responsive to an external stimulus. The alternate layers having a difference in periodic refractive indices give rise to a first reflected wavelength. When in response to the external stimulus, there is a change in the responsive material then the reflected wavelength of the device shifts or changes from the first reflected wavelength to a second reflected wavelength. In different examples the external stimulus may be mechanical stimulus or chemical stimulus or electrical stimulus or thermal stimulus or light stimulus or magnetic stimulus or their combination(s).

CONCLUSION

In this chapter, some recent patents on tunable PCs were reviewed. Mainly, it is found that out of number of parameters which can be used to tune photonic band gaps in PC structure such as refractive indices of materials, thicknesses of material layers and angle of incidence of light on the structure etc. In these patents, researchers have based tuning of photonic band gaps on variation of PC's constituent materials' refractive indices and thicknesses. However, the changing of angle of incidence of light on the structure to tune photonic band gaps is a well known but least explored tuning parameter. The tuning of photonic band gaps by varying angle of incidence of light has its own advantages. Firstly, the original photonic crystal structure will not have to be modified for tuning photonic band gaps, only an external mechanism will be required to change angles of incidence of light on the structure to change transmitted and reflected wavelengths. Secondly, tuning mechanisms, methods and fabrications in patents [8-14] based on varying refractive index and thicknesses of constituent material layers of photonic crystals are complicated and costly. So, a lot of inventions can be done in this field, which will be exploring the idea of tuning photonic band gaps in PCs by changing angles of incidence of light on the structure and its possible combination(s) with other structural tuning method(s) by suitable mechanism, design and idea to get an optimum method or mix of methods for tuning photonic band gaps in photonic crystals.

ACKNOWLEDGMENTS

I am grateful to Sri. Aseem Chauhan, Maj. Gen. K. K. Ohri, Prof. S. T. H. Abidi, Col. Vivek Tripathi, Brig. U.K. Chopra and Prof. N. Ram of Amity University, Uttar Pradesh, Lucknow Campus, India for their constant encouragement and support during this research work. I also offer my thanks to Prof. V.P. Sahi and Prof. Q. Rahman.

REFERENCES

[1] Yablonovitch, E. *Phys. Rev. Lett.,* 1987, vol. 58, 2059-2062.
[2] Yablonovitch, E. *J. Opt. Soc. Am. B,* 1993, *vol.* 10, 283-295.

[3] Joannopoulos, J.D.; Johnson, S.G;. Winn J.N.; Meade, R. D. Photonic Crystals: Molding the Flow of Light, Princeton University Press, Princeton, NJ, 2008.

[4] Banerjee, A. Optical Filtering Characteristics of 1D Photonic Crystals, Lambert Academic Publishing, Germany, 2012.

[5] Banerjee, A. *J. Electromagnet. Wave,* 2010, vol. 24, 1023-1032.

[6] Banerjee, A. *Prog. In Electromagn. Res.,* 2009, vol. 89, 11-22.

[7] Banerjee, A. *Prog. In Electromagn. Res. Lett.,* 2009, vol. 11, 129-137.

[8] Arsenault, A. US20110233476A1, 2011.

[9] Puscasu, I.; Pralle, M.U.; Daly, J.T.; McNeal, M.P. US7825380B2, 2010.

[10] Wilson, C.J.; Sigalas, M.M.; Flory, C.A. US6760514B2, 2004.

[11] Park, W.; Lee, J.-B.; Tinker, M.T.; Schonbrun, E.F. US7283716B2, 2007.

[12] Duncan, W.D.; Hyde, R.A.; Myhrvold, N.P.; Tegreene, C.T.; Weaver, T.A.; Wood, L.L., Jr. Pub. No. US2012/0069417A1, 2012.

[13] Arsenault, A.; Ozin, G. A.; Puzzo, D. P. US7826131, 2010.

[14] Arsenault, A., Ozin, G. A., Puzzo, D. P. US20110164308A1, 2011.

In: Photonic Crystals
Editor: Donald Claus

ISBN: 978-1-62618-193-9

Chapter 5

DESIGN OF HIGH PERFORMANCE TUNABLE REFLECTORS BY USING 1D TERNARY PHOTONIC BAND GAP STRUCTURE

Anirudh Banerjee*
Department of Electronics and Communication,
Amity School of Engineering and Technology, Lucknow,
Amity University, Uttar Pradesh, India

ABSTRACT

This chapter demonstrates a novel way to design high performance tunable reflectors by modifying a 1D binary PBG (Photonic Band Gap) structure to a 1D ternary periodic structure. This approach significantly enhances the incidence angle based tunability of reflection bands in the structure. This incidence angle based tunability of reflection bands can be further enhanced by increasing the thickness of the sandwiched layers in the ternary PBG structure. These 1D ternary PBG structure based high performance tunable reflectors can be substituted in place of 1D binary PBG structure based reflectors where fast and wide reflection band tuning is required.

Keywords: Binary photonic crystal, ternary photonic crystal, photonic band gap, reflection bands, reflectors

---

* Email: anirudhelectronics@yahoo.com.

1. INTRODUCTION

Recently, 1D periodic ternary PBG (Photonic Band Gap) structures have attracted a lot of attention of researchers due to their remarkable superior performances over 1D periodic binary PBG structures in refractometric optical sensing, temperature sensing, and band gap extension [1-3]. These 1D ternary PBG structures are formed by periodic repetition of three different material layers instead of two different materials as in the case of conventional 1D binary PBG structures. Actually, 1D PBG structures are classified into the categories as binary, ternary, quaternary etc., depending on the number of different material layers in one period. 1D binary PBG structure contains two layers in a period; 1D ternary PBG structure contains three layers in a period; 1D quaternary PBG structure contains four layers in a period and so on. Finally, these periods are repeated several times to form PBG structures. These 1D ternary PBG structures not only extend band gap [3], and enhance sensing performance [1-2], but, in this chapter, it is shown that high performance tunable reflectors can be designed by using 1D ternary PBG structures which are superior to tunable reflectors based on 1D binary PBG structures.

2. THEORY

The 1D binary PBG structure [4] consisting of alternate layers of refractive indices and n_2 with thicknesses and d_2 respectively is depicted in figure 1. is the period of the lattice. The binary structure shown in figure 1 can be modified by sandwiching a thin layer of a third material of refractive index and thickness between two layers of the binary structure periodically to form a ternary photonic band gap structure as depicted in figure 2. Now this modified periodic structure (i.e. ternary structure) has alternate layers of refractive indices (first material layer), (second material layer) and (third material layer) with thicknesses: d_1 , d_B and d_2 respectively. is the period of the lattice. The transmission coefficient of this multilayer is given by [1]

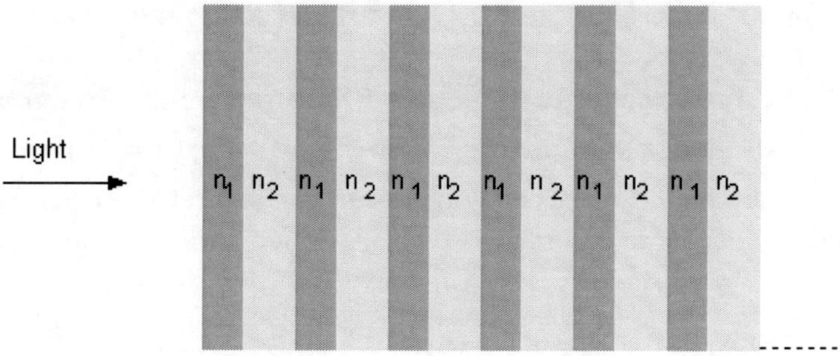

Figure 1. Schematic of 1D binary PBG structure.

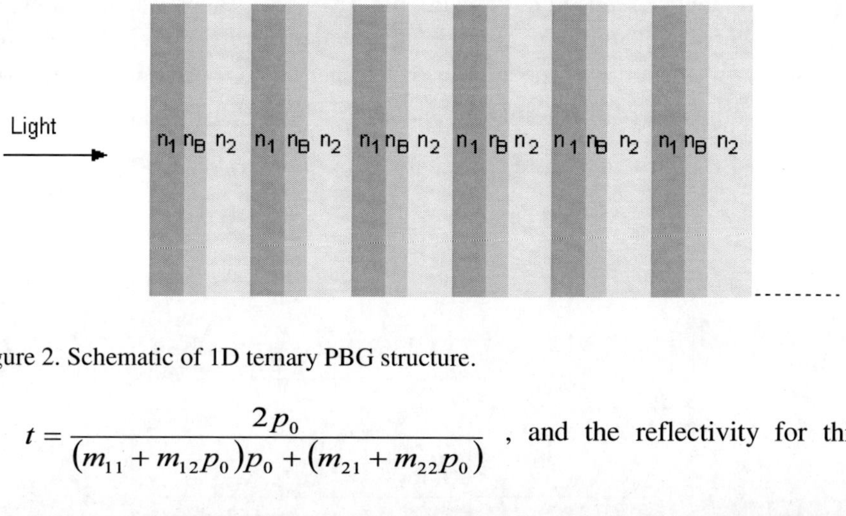

Figure 2. Schematic of 1D ternary PBG structure.

$t = \dfrac{2p_0}{(m_{11} + m_{12}p_0)p_0 + (m_{21} + m_{22}p_0)}$, and the reflectivity for this structure can be written in terms of transmission coefficient as $R = 1 - |t|^2$, where as $(n_0 = 1)$. The reflectivity of the structure, for TM wave can be obtained by using

$$p_1 = \frac{\cos\theta_1}{n_1}, \; p_2 = \frac{\cos\theta_2}{n_B}, \text{ and } p_0 = \cos\theta_0 .$$

3. Results and Discussions

The reflection spectra for 1D binary PBG structure at different angles of incidence $\left(40° \text{ and } 60°\right)$, are shown in figures 3(a) and (b) respectively. Solid curves are for TE waves and dotted curves are for TM waves. The values of various parameters for this 1D binary PBG structure were chosen as $n_1 = 3.3$, $n_2 = 1.46$, $d_1 = 120nm$, and $N = 10$. It was found that the reflection bands shift toward lower wavelengths with increase in angle of incidence of the light on the binary PBG structure. It was observed that the TE reflection bands shift toward lower wavelengths by $117nm$ and $207nm$ at angles of incidence respectively, while, the TM wave transmission peak shifts toward lower wavelengths by at angles of incidence respectively.Figures 3(c) and (d) show the output spectra for 1D ternary PBG structure at different angles of incidence $\left(40° \text{ and } 60°\right)$. The values of various parameters for this 1D ternary PBG structure were chosen as $n_1 = 3.3$, $n_B = 2.04$, $n_2 = 1.46$, $d_1 = 120nm$, $d_B = 30nm$, and $N = 10$.

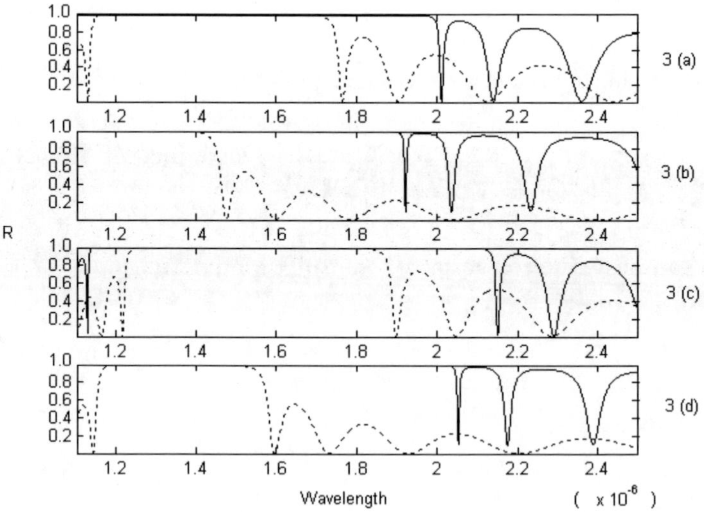

Figure 3(a). Reflection spectra for 1D binary PBG structure at 40° angle of incidence. 3(b). Reflection spectra for 1D binary PBG structure at 60° angle of incidence. 3(c). Reflection spectra for 1D ternary PBG structure at 40° angle of incidence. 3(d). Reflection spectra for 1D ternary PBG structure at 60° angle of incidence.

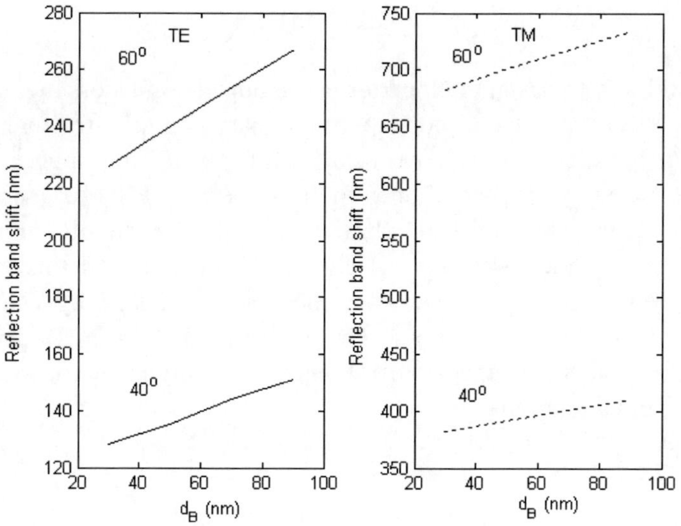

Figure 4. Effect of thickness of sandwiched layers on the reflection band shifts at different angles of incidence.

Here, in this case also, it was found that the reflection bands shift toward lower wavelengths with increase in angle of incidence of the light on the ternary PBG structure. However, the incidence angle sensitive TE wave output spectrum reflection band shift of binary periodic structure was enhanced by $11nm$ and $19nm$ at angles of incidence respectively when the structure was modified by sandwiching a thin layer of third material (with refractive index $n_B = 2.04$), between every two layers, constituting a period of lattice. The incidence angle sensitive TM wave output spectrum reflection band shift of binary periodic structure was enhanced by $18nm$ and $31nm$ at angles of incidence respectively when the structure was modified by sandwiching a thin layer of third material (with refractive index $n_B = 2.04$), between every two layers, constituting a period of lattice. The investigation was then repeated by taking $d_B = 50nm$, and $90nm$. It was found that the incidence angle sensitive TE and TM wave output spectrum reflection band shift further enhances with increase in thickness of the sandwiched layers. Figure 4, shows the plots of incidence angle sensitive output spectrum reflection band shift with the thickness of the sandwiched layer for TE and TM waves at angles of incidence $40°$ and $60°$.

Conclusion

A novel way to design high performance tunable reflectors by modifying a 1D binary PBG (Photonic Band Gap) to a ternary periodic structure (i.e. three material layers constituting a period of lattice) has been suggested. Slight change in angle of incidence of light on the 1D ternary PBG structure causes a sufficiently large reflection band shift, which increases the tuning range and tuning speed of reflectors. The reflection band tunability of this 1D ternary photonic crystal can be further controlled and enhanced by changing the thickness of the sandwiched layer. These high performance 1D ternary PBG structures reflectors are superior to 1D binary PBG structures for wide and fast spectrum tuning requirements.

Acknowledgments

I am grateful to Sri. Aseem Chauhan, Maj. Gen. K. K. Ohri, Prof. S. T. H. Abidi, Col. Vivek Tripathi, Brig. U.K. Chopra and Prof. N. Ram of Amity University, Uttar Pradesh, Lucknow Campus, India for their constant encouragement and support during this research work. I also offer my thanks to Prof. V.P. Sahi and Prof. Q. Rahman.

References

[1] Banerjee, A. *Prog. In Electromagn. Res.,* 2009, vol. 89, 11-22.

[2] Banerjee, A. *Prog. In Electromagn. Res. Lett.,* 2009, vol. 11, 129-137.

[3] Wu C.-J.; Chung, Y.-H.; Syu, B.-J.;Yang, T.-J. *Prog. In Electromagn. Res.* 2010, vol. 102, 81-93.

[4] Banerjee, A. In Photonic Crystals: Fabrication, Band Structure and Applications; Laine, V. E.; Nova Science Publishers: Hauppauge, NY, 2010.

In: Photonic Crystals
Editor: Donald Claus

ISBN: 978-1-62618-193-9

Chapter 6

A NEW MULTILAYER STRUCTURE FOR TEMPERATURE STABLE PERFORMANCE IN INFRARED OPTICAL FILTERING

Anirudh Banerjee*
Department of Electronics and Communication,
Amity School of Engineering and Technology, Lucknow,
Amity University, Uttar Pradesh, India

ABSTRACT

A new multilayer structure is suggested for infrared optical filtering in temperature unstable environments. Large changes in environmental temperature in operating range of filter causes a negligible effect on output spectrum. Due to superior immunity to temperature changes, this proposed new multilayer structure is a superior substitute to traditional 1D binary PBG (Photonic Band Gap) structure where temperature stable performance is required.

1. INTRODUCTION

Periodic media like one-dimensional (1D) binary photonic band gap (PBG) structures [1-4] have attracted great interest of researchers due to their

* Email: anirudhelectronics@yahoo.com.

remarkably useful capabilities to guide and control light propagation. In these 1D binary PBG structures, two different optical materials are periodically arranged. These 1D binary PBG structures lead to the formation of photonic band gaps or stop bands, in which propagation of electromagnetic waves of certain wavelengths are prohibited. These 1D binary PBG structures are currently at the forefront in various optical, technological applications and research. However, a major drawback of these conventional 1D binary PBG structures is that the output spectrum of these structures are highly temperature sensitive, therefore, these PBG structures are practically unsuitable for applications in temperature unstable environments. To overcome this drawback an alternative new multilayer structure [5-6, 8] is suggested in this chapter, the output spectrum of this new multilayer structure is relatively immune to environmental temperature variations. In this new sixteen layer multilayer structure, layers with low refractive index (denoted by *L*) and layers with high refractive index (denoted by *H*) are arranged in the sequence LLLHHLHHLLLHHLHH.

Figure 1. Schematic of the 1D binary PBG structure.

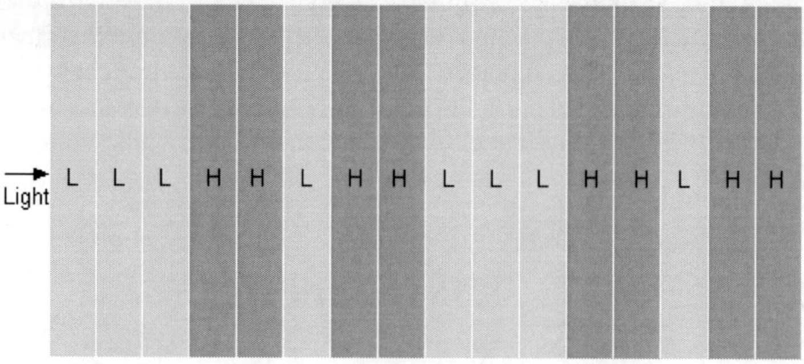

Figure 2. Schematic of the proposed new multilayer structure.

The materials for L and H layers are SiO_2 and Si respectively. On comparison of this proposed new multilayer structure with a sixteen layer 1D binary PBG structure consisting of alternate layers of same low and high refractive indices in the sequence LHLHLHLHLHLHLHLH, it was found that the performance of this new multilayer is many times more immune to environmental temperature variations. Thus, in this chapter this proposed new multilayer structure is suggested as a superior substitute to 1D binary PBG structure for infrared optical filtering applications.

2. STRUCTURES

A sixteen layer 1D binary PBG structure consisting of alternate layers of low and high refractive indices in the sequence LHLHLHLHLHLHLHLH is shown in figure 1. In figure 2, the proposed new multilayer structure is shown. Layers with low refractive indices (denoted by *L*) and layers with high refractive indices (denoted by *H*) are arranged in the LLLHHLHHLLLHHL HH sequence.

3. EFFECT OF TEMPERATURE VARIATION

It is well known that the thickness and refractive index of a medium is changed during temperature rise due to thermal expansion effect and thermo-optical effect [4]. Using the matrix formulation method the transmittance through 1D binary PBG structure and this proposed multilayer structure can be calculated [7-8]. In the previous work [8], it was observed for a 1D binary PBG structure with $n_L = 1.46(SiO_2)$, $n_H = 3.3(Si)$, and and $N = 8$, the transmission peak shifts by towards higher wavelengths with per degree rise in temperature. While, in the case of this proposed multilayer structure, with the same structural parameters the transmission peak shifts by towards higher wavelengths with per degree rise in temperature. So, the temperature sensitive transmission peak shift of (Si/SiO_2) periodic structure was reduced by seven times per degree when the structure was modified to form new multilayer structure. Further, in that work [8], for a 1D binary PBG structure with $n_L = 1.46(SiO_2)$, $n_H = 3.3(Si)$, $d_L = 300nm$, and it was found that the transmission peak shifts by towards higher wavelengths with per degree rise in temperature. While in case of this proposed new multilayer structure, with same structural parameters, it was found that transmission peak shifts by towards higher wavelengths with per degree rise in temperature. Thus, the temperature sensitive transmission peak shift of (Si/SiO_2) periodic structure was reduced to half when the structure was modified to form new multilayer structure. In addition to this, it is also observed that this new multilayer structure exhibits narrow transmission peaks in the output spectra as compared to 1D binary PBG structure. Thus, this new multilayer structure is also attractive for narrowband infrared optical filtering applications in WDM (Wavelength Division Multiplexing). This new multilayer structure demonstrates a superior optical filtering performance which is relatively more immune to environmental temperature variations.

CONCLUSION

A new multilayer structure with temperature stable performance is suggested for optical filtering applications. On comparison with 1D binary PBG structures it is found that the large changes in environmental temperature in this new multilayer structure causes a sufficiently low transmission peak shift, which is negligible and will hardly effect the filtering performance of the system in which this structure will be used. The size, weight and cost of the 1D binary PBG structure and the proposed new multilayer structure will be same due to the presence of same number of layers of the same two materials. However, due to superior immunity to temperature changes and narrow peak generation in output spectrum this proposed new multilayer structure is a superior substitute to 1D binary PBG structure for infrared optical filtering applications where temperature stable performance is required.

ACKNOWLEDGMENTS

I am grateful to Sri. Aseem Chauhan, Maj. Gen. K. K. Ohri, Prof. S. T. H. Abidi, Col. Vivek Tripathi, Brig. U.K. Chopra and Prof. N. Ram of Amity University, Uttar Pradesh, Lucknow Campus, India for their constant encouragement and support during this research work. I also offer my thanks to Prof. V.P. Sahi and Prof. Q. Rahman.

REFERENCES

[1] Yablonovitch, *E. Phys. Rev. Lett.,* 1987, vol. 58, 2059-2062.

[2] Joannopoulos, J. D.; Johnson, S.G.; Winn, J. N.; Meade, R. D. Photonic Crystals: Molding the Flow of Light; Princeton University Press: Princeton, NJ, 2008.

[3] Banerjee, A. In Photonic Crystals: Fabrication, Band Structure and Applications; Laine, V. E.; Nova Science Publishers: Hauppauge, NY, 2010.

[4] Kai, T.; Wei, C.W-; -hui, Y.G; Zhi-quan, L. *Optoelec. Lett.,* 2007, vol. 3, 0444-0447.

[5] Banerjee, A. In Horizons in Computer Science Research; Clary, T.S.; Nova Science Publishers: Hauppauge, NY, 2011, Vol. 6.

[6] Banerjee, A. In Optical Lattices: Structures, Atoms and Solitons; Fuentes, B. J.; Nova Science Publishers: Hauppauge, NY, 2011.

[7] Born, M.; Wolf, E. Principles of Optics; Cambridge University Press, UK, 1980.

[8] Banerjee, A. In Advances in LASER and Optics Research; Arkin, W.T.; Nova Science Publishers: Hauppauge, NY, 2011, Vol. 9.

INDEX

D

E

F

G

H

I

L

M

N

O

P

S

T

U